Science and Anti-Science

Science and Anti-Science

GERALD HOLTON

HARVARD UNIVERSITY PRESS
Cambridge, Massachusetts
London, England

PRINTED IN THE UNITED STATES OF AMERICA
Second printing, 1994

First Harvard University Press paperback edition, 1994

Library of Congress Cataloging in Publication Data

Holton, Gerald James.
Science and anti-science / Gerald Holton.
p. cm.
Includes bibliographical references and index.
ISBN 0-674-79298-X (alk. paper) (cloth)
ISBN 0-674-79299-8 (pbk.)
1. Science—Philosophy. 2. Science—History. I. Title.
Q175.H7748 1993
501—dc20 93-272
CIP

To Nina

Contents

Preface

What are the earmarks of good science? What goal—if any—looms as the proper end of all scientific activity? What legitimating authority may scientists claim?

These old questions, to which each era attempts its own response, are being debated today with renewed vigor. For this book I have selected answers that have emerged mostly in our century and primarily from the words and actions of scientists and scientist-philosophers. As readers of my previous books would expect, I aim to understand these words and actions not in the abstract, but in the natural setting of specific historic cases.

Thus the first chapter traces how the nineteenth-century empiricist view of what good science should be—chiefly the version represented in Ernst Mach's writings—came to influence, often in quite indirect and transmuted form, the thoughts of twentieth-century scientists and philosophers such as Jacques Loeb, B. F. Skinner, Philipp Frank, P. W. Bridgman, W. V. Quine, and some of their colleagues. The chapters that follow deal similarly with controversies and rhetoric that illuminate positions on the proper use, goals, and legitimacy of science, as expressed by such seminal figures as Albert Einstein, Max Planck, and Niels Bohr, but also by less-known ones such as Joseph Petzoldt and Walter Kaufmann. Because it has become clear that the two standard models for the pursuit of research, which generally are traced back to Newton and Bacon, respectively, are no longer adequate to the needs of our time, Chapter 4 is devoted to the rise of a third, additional solution, one that arguably has roots in Thomas Jefferson's approach to science.

Increasingly during the last half of this century, voices are heard from various directions that "good science" is inherently an oxymoron, that science as we have known it is ultimately either self-destructive (Oswald Spengler's position) or disruptive of the social equilibrium (as, for example, in the passages cited from the writings of Václav Havel). Thus the last two chapters focus on how to understand two confrontations: one is the conflict between the view that the sciences are by their nature subject to eventual decay and the contrary view that the sciences are destined to merge into one coherent body of understanding for all natural phenomena; the other is the more public battle between practitioners of science and opponents who champion "alternative" or anti-science, particularly the form that is based in a world picture within which anti-science is an organic part of a politically ambitious movement. Here, as throughout the book, there is evident interplay between the interests of scholarship and the turbulent course of public debate; my hope is that advancing the former provides some clarification for the latter.

I gladly acknowledge support for research on portions of this book from the Andrew W. Mellon Foundation, and once more wish to record my gratitude for the expert help of Ms. Joan Laws in all the tasks needed to convert ideas into printed pages.

Science and Anti-Science

1

Ernst Mach and the Fortunes of Positivism

Between 1910 and 1914, the Nobel Committee in Stockholm received a number of letters and petitions from scientists nominating Ernst Mach for the Nobel Prize in physics. Among these, H. A. Lorentz praised Mach's "beautiful works," especially in acoustics and optics, which indeed have not lost their luster to this day, and he added that "all physicists" know Mach's historical and methodological books, and "a large number of physicists honor him as a master and as a guide of their thoughts." (A few years later, in his obituary for Mach in 1916, Albert Einstein said it more strikingly: "I believe even that those who consider themselves as opponents of Mach are hardly aware of how much of Mach's way of thinking they imbibed, so to speak, with their mother's milk.") Ferdinand Braun's nominating letter indicated that while the Nobel Prize might soon have to be awarded for the new theory of space and time, it should first go to Mach, as an early advocate of ideas along the same lines as well as a major experimental physicist; but Braun, too, insisted on Mach's wider influence through "his clear, profound historical-physical studies" and philosophical clarifications.[1]

As is well known, only a few years before those letters were sent to Stockholm, Einstein—who later confessed in his *Autobiography* that Mach's *Mechanik* had "exercised a profound influence" upon him, and that Mach's example of critical reasoning had been a requirement for his discovering the key to relativity—signed one of his letters to Mach "Your admiring student."[2] Similarly, the next generation of physicists, which struggled with the problems of the new quantum mechanics (e.g., Werner Heisenberg and Wolfgang Pauli), also found in Mach guidance for their thoughts.

At each point in the history of science, we find a few individuals who are thought by their contemporaries to be pointing toward new answers for old questions concerning the proper task of scientific practice and the place of science in culture. From the 1880s to the first decades of the twentieth century, Mach was one of those few. At least among scientists, he was recognized as one of the most effective fighters in the empiricist challenge to notions implying "absolutes" that had permeated nineteenth-century science (e.g., absolutes of space, time, substance, vital force). Among philosophers, Mach was admired or attacked for his vigorously held empiricist vision of science, of which perhaps the most essential point was caught in a succinct paragraph by the philosopher Moritz Schlick:

> Mach was a physicist, a physiologist, also a psychologist, and his philosophy . . . arose from the wish to find a principal point of view to which he could cling in any research, one which he would not have to change when going from the field of physics to that of physiology or psychology. Such a firm point of view he reached by going back to what is given before all scientific research, namely, the world of sensations. . . . Since all our testimony concerning the so-called external world relies only on sensations, Mach held that we can and must take these sensations and complexes of sensations to be the sole content of those testimonies, and therefore that there is no need to assume in addition an unknown reality hidden behind the sensations. With that, the existence of *Dinge an sich* is removed as an unjustified and unnecessary assumption. A body, a physical object, is nothing else than a complex, a more or less firm pattern of sensations, i.e., of colors, sounds, sensations of heat, of pressure, etc.[3]

Even though Mach repeatedly disavowed that he was proposing a systematic philosophy, he took every opportunity to ensure that his influence would extend far beyond physics, just as he was intent that it would go beyond the borders of his homeland. And indeed it turned out that his teachings lent themselves—more often through the Machian spirit than through direct transmission of raw positivist statements—to adoption or adaptation by many throughout Europe and America who longed for modernism across a great spectrum of intellectual endeavors and who were infected by the unsuppressible minority view that rejected blatantly metaphysical and hierarchical systems in favor of a unified, empirically based world conception. After Mach assumed his professorship in experimental physics at the University of Prague in 1867, there developed a far-flung network of admirers and critics of his ideas that, within a few decades, made him one of the seminal figures in the shaping of the modern world view.[4] His work came to be read, debated, and used not only by physicists but also by major thinkers in mathematics, logic, biology, physiology, psychology, economics, history and philosophy of science, jurisprudence, sociology, anthropology, literature, architecture, and education.[5]

At first slowly and then more and more rapidly, Mach's teachings, often in considerably modified versions, came to be built into the thinking of scholars throughout Europe and, as we shall see, especially in the country that, to his regret, he was never able to visit but had called "the land of my deepest longing," the United States of America.[6] Indeed, the argument can be made, and will be supported here by exemplary cases, that in the long run there was no more fruitful soil for the development and transformation of Mach's ideas than the United States, the country traditionally open to empiricism and pragmatism. The readiness of American scholars in the nineteenth century to be hospitable to some versions of European positivism or empiricism has been discussed, for example, in connection with J. B. Stallo and C. S. Peirce. They did not lack for independent ideas, and they frequently expressed frank disagreements; on that point, one need only follow the original thoughts in Stallo's *Concepts and Theories in Modern Physics* or read the scathing passages in Peirce's review of Mach's *Mechanik.* But America was ready for Mach. There is some poetic justice in this. As we know from Mach's various autobiographies, the Benedictine fathers of his gymnasium thought him to be unteachable and without talent; in turn, the young man felt so oppressed by the regime in Austria-Hungary that he prepared himself to emigrate to America.[7]

To be sure, Mach hardly fitted into the mainstream of Austria-Hungary. He was a freethinker (a fact that later held up his appointment to the professorship in Vienna); politically most nearly identifiable with socialism of the Austrian type; an active fighter against fanatical nationalism and anti-Semitism (the latter even certified by a Prague police report); and a tireless propagandist for "a point of view free from metaphysics, as a product of the general evolution of culture."[8]

In the methodology of science, too, Mach at first was an outsider. One of his earliest essays, on a new, instrumentalist basis for defining mass, written in 1866 and already indicating his powerful methodological point of view, was returned as unpublishable by J. C. Poggendorff's *Annalen der Physik.* It is difficult to realize today how shaky and dogmatic the fundamentals of the physical sciences were prior to the clarifying work in which Mach participated in the last third of the nineteenth century, when some German textbooks in physics still implied that the meaning of concepts was to be sought on a higher, metaphysical plane. What made the difference eventually, in physics but in other fields too, was in good part that philosophically minded young scientists in many countries, in their student years or soon after, and often in reading clubs that they initiated, chanced upon and became fascinated with the writings of Mach and related works. Among these were works by Henri Poincaré, sixteen years younger than Mach, who in direct ways expressed his debt to Mach; and by Pierre Duhem, who wrote to Mach on 10 August 1909: "Permit me to call myself your disciple."[9] These two, together with Hermann von Helmholtz, Gustav R. Kirchhoff, Wilhelm Ostwald, Richard Avenarius, Ernst Haeckel, J. B. Stallo, Karl Pearson, and others of that general cast of mind, were the chief authors of the eagerly read tribal books for guiding thought into the new age.

Paul Carus (1852–1919)

Not only scientists and scholars but a variety of interested laymen were attracted to Mach's ideas. In the early phase of the introduction of Mach to America, the crucial and insufficiently recognized intermediary was Paul Carus, editor of the journals the *Open Court* and the *Monist* as well as of the parent firm, the Open Court Publishing Company. Born in Germany and with a doctorate from the University of Tübingen, Carus was an amateur philosopher and indefatigable author who sought to develop an agnostic, monistic, and evolutionist world view. He engaged in a massive, mostly unpublished correspondence with Ernst Mach for almost three decades—one of the largest Mach had with anyone—of which many letters have survived. Through their letters one can get a sense of Mach's interaction with contemporaries who exhibited interest in his ideas.[10]

Over the years Carus's publishing enterprise, located in the small prairie town of La Salle, Illinois, saw to it that as many as possible of Mach's works would be made available in English; this included a large number of articles and fifteen books (in original editions, reprintings, etc.). Mach's *Popular Scientific Lectures* (1895) appeared in English even before the German edition (1896), as did three chapters of *Erkenntnis und Irrtum,* later gathered by Carus into the little book *Space and Geometry.* Mach was an enthusiastic collaborator in this constant output by his American publisher, saying in a letter to Carus of 26 August 1890, "It is particularly important for me that the *Analysis of Sensations* appear in America," and on 20 March 1894, "I lay *particular* value on writing for the American circle of readers." Similarly, on 11 August 1889 he accepted Carus's proposal that his article in the very first issue (1890) of the *Monist* (entitled, again in line with a suggestion by Carus, "The Analysis of Sensations—Anti-Metaphysical") would carry the introductory note: "The time seems ripe for the overthrow of all metaphysical philosophies. I contribute this article to your magazine in the confidence that America is the place where the new views will be most developed. E.M."

The hope the men shared, that these publications would attract an ever-widening circle of American readers to Mach's ideas, soon began to be fulfilled; and even though the most prominent scholars among them could of course read Mach's works in the original German editions, they tended to cite these English translations. As Mach noted with satisfaction, his *Mechanik* had a much larger distribution in the English version brought out by Carus than in the original German one.[11]

Many of the bare facts in the relation between Carus and Mach have been known for some time.[12] What has been missing, but is needed to understand how Carus in this then-unlikely outpost could become Mach's first missionary in the United States, is a more detailed, sympathetic understanding of what these two men meant to each other, as well as a sense of how the collaboration of this odd pair amounted to an act of inspired symbiosis. Such a treatment will have to be given elsewhere; suffice it to note here that Carus had read Mach's *Mechanik* with greatest interest when it appeared in 1883 and later wrote, "I at once recognized in him a kindred spirit." Indeed, the English translation of the *Mechanik* was one of the first projects of the fledgling company, Mach assuring Carus, "I will be very glad to work over the English text."[13] Moreover, during Mach's lifetime, apart from two early pieces in the *Philosophical Magazine* (London; 1865, 1866), all twenty articles by Mach that appeared in English, whether translated from published works or from manuscripts, appeared in the *Open Court* or the *Monist.* The parent company was also responsible for all of his books in English translation.

Carus clearly revered Mach, even if he occasionally, in letters, editorials, and articles, expressed reservations on specific points. He saw himself as a fellow intellectual whose "admiration for Professor Mach cannot be less than that of his most ardent disciple and follower." In return, Mach's letters demonstrated his pleasure and respect. For example, he wrote to Carus on 26 January 1890 that he was delighted with Carus's new book *Fundamental Problems* and added, "Your motto . . . 'positive science' is one with which I am in full agreement. In general, your monistic conception is very sympathetic to me, and I find many points of contact with my own considerations."[14]

One would expect Mach to reach out to Carus in this way. He was not one to leave to chance any opportunity for increasing his influence. But there can be little doubt that the sentiment behind Mach's published and private expressions of gratitude to Carus was genuine. And of course he had uncommonly good reason to feel that way. The labors of Carus's firm were putting him in touch with "the American public" and made his work "become international."[15] Carus made at least two pilgrimages to Mach (1893, 1907) and planned at least one additional visit (1913). Mach may also have perceived the La Salle operation in the context of his long-standing sympathy for America. Like so many Europeans, he may have had a somewhat romantic attitude on this score; thus, among the books on America that Carus undertook to send him from time to time, there were some on American Indians, in whom Mach was also interested through anthropological studies. Then, too, the New World continued to beckon the Mach family. On various occasions his son Ludwig planned to emigrate there; Mach himself continued to hope for at least a visit, although nothing came of it. And Mach had not only admirers in the United States but also at least one family member, a cousin—William Lang—in Chicago.[16]

In short, Mach's appearance on the American scene was so massive and successful in good part because of the multiple bonds between Mach and Carus. In addition to their personal friendship, they perceived each other to be kindred spirits indeed, as well as beleaguered fellow outsiders who were fighting for a common vision of the modern scientific world conception while all around them, as Mach put it, "This is the time of anti-modernism."[17]

William James (1842–1910)

The first major American scientist on whom Mach's work had documentable direct influences—and who did not have to wait for the English translations—was William James. Only a few years younger than Mach, James was by education and interests well matched with Mach: he had originally trained in physical science and medicine; had traveled widely in Europe, including a brief but important period as a student in Berlin, Heidelberg, and Dresden; and had served as instructor of anatomy and physiology and, finally, as professor both of philosophy and of psychology at Harvard University. His profound and influential *Principles of Psychology* (1890) and *Pragmatism* (1907) established his reputation as one of the seminal thinkers of his day in America, one of the few American philosophers read widely in Europe.

His philosophy of pragmatism, developed in the first instance as a way out of a personal struggle that has been called James's "Kant crisis," overlapped with Machian positions in many ways, for example, in finding the meaning of ideas in the sensations that may be expected from their realization. As early as 1875 he was reacting against the previous generation of what he called "the Heaven-scaling Titans" of Germany, against whom he named Gustav Theodor Fechner, Helmholtz, and Mach among models of the new breed whose "detachment of mind is very healthy" from either "theologic or anti-theologic bias." James was also impressed by Mach's experimental results, to which he referred in many places in his writings; in addition, as Ralph B. Perry noted, "From Mach, James had learned something of what he knew about the history of science, and he had readily accepted his view of the biological [evolutionary] and economic function of scientific concepts."[18]

James himself put his relation with Mach most perceptively when he wrote on 27 June 1902, in response to Mach's request that he accept the dedication of the new, third edition of *Populärwissenschaftliche Vorlesungen:*

> I suppose that sympathies are usually reciprocal, and that just as I have taken such delight in the whole tone and temper of your thoughts, so you have found something in the tone of my writings, imperfect though they are, which has pleased you . . . I trust . . . that you and I may yet read many further productions by each other and contribute jointly to the establishment of the truly philosophical way of thinking—which I believe to be, on the whole, *our* way![19]

A few months later, having received Mach's new book with the dedication in "Sympathie und Hochachtung" ("sympathy and deep respect"), James wrote on 19 November 1902 of his attempt to give his students at Harvard "a description of the construction of the world built up of 'pure experiences' related to each other in various ways, which are also definite experiences in their turn."

In using the word *sympathy,* both men pointed to the heart of the troublesome concept "influence." It is not enough to find passages in which one person echoes another. What counts far more is the development of a state of elective—but also selective—affinity, as was the case for Mach and James.[20]

It is of course no accident that psychology was the first scientific field to feel Mach's impact. His own researches in experimental psychology and psychophysics during much of his life made it likely that this part of his work would be read attentively among psychologists in America.[21] They were also open to his ideas in and beyond psychology because of the native empiricist tradition in the United States in psychology and philosophy, fields that at that time were not yet clearly disaggregated. To preview a subsequent development, we may note that even in 1930, when Moritz Schlick returned to Vienna from a missionizing visit to the United States and gave at the newly founded Ernst-Mach-Verein a report entitled "The Scientific World Conception of the United States," he singled out one field above all others, saying, "The great respect for empirical psychology provides a favorable ground for the scientific world conception" there. And Herbert Feigl, one of Schlick's favorite students, having gone to America in 1930 as "the first 'propagandist' of our outlook," returned to Vienna with the news that he considered the behaviorist psychologists in the United States to be among "the closest allies our movement acquired in the United States."[22]

That alliance noted in the 1930s had been prepared, as it were, in part by the contacts between James and Mach, starting in the 1870s.[23] In 1882, when both James and Mach were still only at the threshold of becoming widely known, James came to Prague during one of his European study tours and wrote to ask Mach for an interview, noting that he, James, was "very familiar" with his writings. Mach (whose English was quite good) had also read some of James's studies, and the two had a glorious meeting in Prague on 2 November 1882, in which James was overwhelmed by Mach's intellectual power.[24] There followed years of interactions through citations of each other's publications as well as in their correspondence.[25] The originals of six of Mach's letters to James are at the Houghton Library at Harvard. On 29 January 1884 Mach announces to James that the latter's "schönen Versuche" ("beautiful experiments") will be taken into account in his new book; on 17 October 1890 he calls the just-received copy of *Principles of Psychology* "very beautiful and interesting." On 10 June 1902 he thanks James for "so many hours filled with instruction and pleasure." In his well-known letter of 28 June 1907, Mach confirms that in his way of thinking he stands close to pragmatism, although without ever having used that term. And on 6 May 1909 he acknowledges that to James's books, of which he had now a substantial number, "I may thank a large number of new points of view."[26]

In addition to the citations and quotations in their publications and the indications in the correspondence, a third essentially instrumentalist evidence of interaction exists in the form of annotations in published works. William James's extensive collection of books was partly dispersed after his death, but a large number were preserved in Houghton Library at Harvard. In many of these books he had entered annotations in the form of marginalia, underlinings, queries, summaries, and so forth. The authors range from Descartes to Stallo and Bernard Brentano.[27]

It is clear from references in James's publications that he had access to earlier editions and copies other than those that have survived; but an indication of the care with which he read Mach's works can be seen from James's annotations of his extant copies of *Erkenntnis* and *Analyse.* A quick scanning of the first yields thirty-two underlinings (including sentences and parts of paragraphs) and thirteen marginal comments (ranging from "W. J.," to indicate a similarity with his own views, to substantial comments indicating agreement or disagreement).[28]

Similarly, in James's copy of *Analyse* there are thirteen underlinings or other markings in the first thirty-eight pages, and more than two dozen annotations. Moreover, in both books James made indexes of his own inside the back covers, with special references to those passages that showed similarities with his position or current interest. Reading these markings, annotations, and indexes gives substance to James's comment, in his letter to Mach acknowledging receipt of *Erkenntnis,* that he will "devour it greedily." As Judith Ryan has shown, in the case of Mach's *Analyse* "James was clearly combing [this book] for pragmatic leanings that might confirm his own belief in the value of everyday reasoning." Examination of James's copy also supports her argument that Mach's treatise, in the 1886 edition, was the "hidden link between James' 1884 essay ["The Stream of Thought"] and the 1890 chapter in his *Principles of Psychology.*"[29]

In terms of James's earlier intellectual formation, even more significant may be the annotations in his copy (also preserved at Houghton Library) of Mach's *Mechanik* in the original edition of 1883, for which James again prepared his own supplementary index of important ideas. This is the book that starts with the famous challenge: "This work is not a text to drill the theorems of mechanics. Rather, its intention is one of enlightenment—or, to put it still more plainly, an anti-metaphysical one." Hence it is by no means easy to grasp—being simultaneously a study in the history of science, a detailed analysis of topics in mechanics, a tract on how to make one's ideas clear, and a sequel to certain eighteenth-century Enlightenment treatises. But James's copy shows that he mastered it; judging by his annotations, he appears to have been most interested in Mach's discussion of Newton's views on time, space, and causality and in what James's index calls the "empirical character" of concepts such as that of equilibrium, for which James searched the work carefully, finding entries for twelve pages.

Other authors have treated aspects of the correspondence between Mach and James, their agreements and occasional disagreements.[30] But James's copies of Mach's book graphically demonstrate the intense impression they made on him during the period in which he was engaged in writing his own major works.

Jacques Loeb (1859–1924)

After the death of William James in 1910, American thinking in psychology was perhaps most influenced by John B. Watson, who may be considered to have launched the school of behaviorism with his article "Psychology as the Behaviorist Sees It" (1913). This school soon encompassed three of the behavioral (or, more properly, neobehavioral) psychologists who in their respective work periods were arguably the most predominant experimental psychologists in the United States: Edward C. Tolman, Clark L. Hull, and B. F. Skinner.[31] Like Watson himself, each had a large, acknowledged debt to the Machian philosophy of science. But the lines of descent from Ernst Mach to these men are so variegated, and the strands in the growing network of intellectual alliances, oppositions, and personal relationships are multiplying so rapidly as we come to the more recent period, that we must make a brief detour to mention one of Watson's important teachers in the United States, the almost fanatical physicalist interpreter of animal behavior Jacques Loeb.

Loeb was born in the Rhineland in 1859, educated at Strassburg and elsewhere in one of the main traditions of German physiology of the time, came to the United States in 1891, taught at various universities, and finally joined the Rockefeller Institute in New York in 1910. His best-known work, in the period roughly from 1890 to 1915, was on artificial parthenogenesis and tropism; his most powerful book, whose message was clear in its very title, was *The Mechanistic Conception of Life,* expanded from an invited address given in 1906 at the first congress of the International Monist League.[32]

In 1887 Loeb had been troubled by fundamental questions both in biology and about his duties as a scientist. To clarify his thoughts, he initiated a correspondence with Ernst Mach at Prague. He wrote to Mach: "Your *Analysis of Sensations* and your *History of Mechanics* are the sources from which I draw the inspiration and energy to work," and he cited particularly the first chapter of the former book, which is entitled simply "Antimetaphysical," as expressing ideas Loeb acknowledged to be "scientifically and ethically the base on which I stand, and on which, in my opinion, the scientist must stand."[33]

As was his habit, Mach generously responded to and nurtured a promising link with a new disciple. The correspondence continued for over a decade, with Loeb calling Mach "a mentor" and "my teacher." Loeb's initial scientific program was essentially derived from Mach (and in part from Mach's closest friend, the engineer and social reformer Joseph Popper-Lynkeus). This included acceptance of "Mach's attack on 'metaphysical' tendencies in science, his faith in the ethical values inherent in research, and his belief in the fundamental unity of science and technology."[34]

Loeb's adherence to such ideals is also evidenced by his associating himself, as one of its thirty-three signers, with a strange and revealing document. It is the public manifesto, issued sometime between late 1911 and summer 1912 on behalf of a newly emerging Gesellschaft für positivistische Philosophie, under the title

Aufruf!

Eine umfassende Weltanschauung auf Grund des Tatsachenstoffes vorzubereiten, den die Einzelwissenschaften aufgehäuft haben, und die Ansätze dazu zunächst unter den Forschern selbst zu verbreiten, ist ein immer dringenderes Bedürfnis vor allem für die Wissenschaft geworden, dann aber auch für unsere Zeit überhaupt, die dadurch erst erwerben wird, was wir besitzen.

Doch nur durch gemeinsame Arbeit vieler kann das erreicht werden. Darum rufen wir alle philosophisch interessierten Forscher, auf welchen wissenschaftlichen Gebieten sie auch betätigt sein mögen, und alle Philosophen im engeren Sinne, die zu haltbaren Lehren nur durch eindringendes Studium der Tatsachen der Erfahrung selbst zu gelangen hoffen, zum Beitritt zu einer Gesellschaft für positivistische Philosophie auf. Sie soll den Zweck haben, alle Wissenschaften untereinander in lebendige Verbindung zu setzen, überall die vereinheitlichenden Begriffe zu entwickeln und so zu einer widerspruchsfreien Gesamtauffassung vorzudringen.

Um nähere Auskunft wende man sich an den mitunterzeichneten Herrn Dozent M. H. Baege, Friedrichshagen b. Berlin, Waldowstraße 23.

E. Dietzgen, Fabrikbesitzer u. philos. Schriftsteller Bensheim.	**Prof. Dr. Einstein,** Prag.	**Prof. Dr. Forel** Yvorne.
Prof. Dr. Föppl, München.	**Prof. Dr. S. Freud,** Wien.	**Prof. Dr. Helm,** Geh. Hofrat, Dresden.
Prof. Dr. Hilbert, Geh. Reg.-Rat, Göttingen.	**Prof. Dr. Jensen,** Göttingen.	**Prof. Dr. Jerusalem,** Wien.
Prof. Dr. Kammerer, Geh. Reg.-Rat, Charlottenburg.	**Prof. Dr. B. Kern,** Obergeneralarzt u. Inspekteur der II. Sanitäts-Inspektion, Berlin.	**Prof. Dr. F. Klein,** Geh. Reg.-Rat, Göttingen.
Prof. Dr. Lamprecht, Geh. Hofrat, Leipzig.	**Prof. Dr. v. Liszt,** Geh. Justizrat, Berlin.	**Prof. Dr. Loeb,** Rockefeller-Institute, New-York.
Prof. Dr. E. Mach, Hofrat, Wien.	**Prof. Dr. G. E. Müller,** Geh. Reg.-Rat, Göttingen.	**Dr. Müller-Lyer,** München.
Josef Popper, Ingenieur, Wien.	**Prof. Dr. Potonié,** Königl. Landesgeologe, Berlin.	**Prof. Dr. Rhumbler,** Hann.-Münden.
Prof. Dr. Ribbert, Geh. Medizinalrat, Bonn.	**Prof. Dr. Roux,** Geh. Medizinalrat, Halle a. S.	**Prof. Dr. F. C. S. Schiller,** Corpus Christi College, Oxford.
Prof. Dr. Schuppe, Geh. Reg.-Rat, Breslau.	**Prof. Dr. Ritter v. Seeliger,** München.	**Prof. Dr. Tönnies,** Kiel.
Prof. Dr. Verworn, Bonn.	**Prof. Dr. Wernicke,** Oberrealschuldirektor u. Privat-Dozent, Braunschweig.	**Prof. Dr. Wiener,** Geh. Hofrat, Leipzig.
	Prof. Dr. Th. Ziehen, Geh. Medizinalrat, Wiesbaden.	

M. H. Baege, Dozent d. Freien Hochschule Berlin Friedrichshagen.	**Prof. Dr. Petzoldt,** Oberlehrer u. Priv.-Dozent, Spandau.

Figure 1. Appeal for the formation of the Gesellschaft für positivistische Philosophie. (Courtesy of Wilhelm-Ostwald-Archiv, Deutsche Akademie der Wissenschaften zu Berlin.)

printed in bold, large type: *"Aufruf!"*[35] The document (Figure 1) deserves more attention than it has received, not least because it is in some respect a striking preview of the core tenets of another, more famous group publication, the seminal manifesto of the Vienna Circle that would be issued in 1929.

The text of the *Aufruf,* as much a "call to arms" as an appeal for support, runs as follows (in my translation):

> *Appeal!*
>
> To bring forth a comprehensive Weltanschauung, based on the factual material that has been accumulated by the separate sciences, is an ever more urgent need; this is true first of all for science [*Wissenschaft*] itself, but also for our era as such, which will only thereby have earned what we now own.
>
> But this claim can be achieved only through the common labors of many. Therefore we call upon all philosophically interested researchers—no matter in which scientific fields they may be active—and upon all philosophers in the narrow sense of the term whose expectation is to reach by themselves valid knowledge only through the penetrating study of the facts of experience, to join a Society for Positivistic Philosophy. The Society shall have the purpose of establishing lively connections among all the sciences, of developing everywhere the unifying ideas [*vereinheitlichende Begriffe*], and thus press forward toward a contradiction-free unitary conception [*Gesamtauffassung*].

The task of launching this appeal had been put in the hands of two men. One was Mach's most beloved acolyte, the Spandau schoolteacher Joseph Petzoldt, who soon resurfaced as editor of the newly founded *Zeitschrift für positivistische Philosophie* (issued as "Organ der Gesellschaft für positivistische Philosophie") and the author of its first article, an elaboration of the call to arms under the title "Positivistische Philosophie" that pointed to Mach and Kirchhoff as incarnations of the ideal. The other man was M. H. Baege, docent in Berlin, who would soon be referred to in the *Zeitschrift* as its publisher.[36]

The long list of signatories of the *Aufruf* formed an impressive, widely dispersed group of individuals who, despite individual differences, felt they could agree on these central beliefs. They included, in addition to Mach himself, such figures as Albert Einstein (Prague), August Föppl (Munich), Sigmund Freud (Vienna), Georg Helm (Dresden), David Hilbert (Göttingen), Wilhelm Jerusalem (Vienna), Felix Klein (Göttingen), Joseph Popper [-Lynkeus] (Vienna), F. C. S. Schiller (Oxford), and Ferdinand Tönnies (Kiel). It was this emerging "thought collective" that "Prof. Dr. Loeb" of New York chose to join as a fellow signer.[37]

B. F. Skinner (1904–1990)

The detailed connections between Mach and Loeb, and between them and the later behaviorists in the United States, are inviting research topics. But we can focus here only on the propagation of Mach's effect in the formation of one major psychologist of the recent period, Burrhus F. Skinner. Born in 1904, he was to his death in 1990 undoubtedly the most direct and self-confessed disciple of Ernst Mach among native-born American scientists in this century. As he stated in his autobiography, *The Shaping of a Behaviorist,* he recalled only two science books he had read as an undergraduate: Loeb's *Comparative Physiology of the Brain and Comparative Psychology* and *The Organism as a Whole,* with their largely positivistic approach to the study of the behavior of animals. When Skinner came to Harvard University to do his graduate work in 1928, his thesis supervisor, in whose laboratory he remained for five years, was the physiologist W. J. Crozier. It is not accidental that Crozier's own teacher had been Jacques Loeb. Indeed, "it was this ultra-positivistic form of Loebian biology that Skinner encountered at Harvard."[38]

But before Skinner was ready to choose either his research topic or Crozier's laboratory, while still in his preparatory courses at Harvard, another push in the same direction came while he was taking a course on the history of science given by George Sarton—who regarded himself as a Duhemian positivist—and the physiologist Lawrence J. Henderson. There Skinner was assigned the reading of Mach's *Mechanics.* It had a permanent effect on him. In an interview on 8 June 1988 Skinner stated to me categorically: "I was totally influenced by Mach via George Sarton's course, and quickly bought Mach's books, *Science of Mechanics* and *Knowledge and*

Error. My Ph.D. thesis was published[39] as 'The Concept of the Reflex in the Description of Behavior.'" In that interview he delighted in adding that a fellow graduate student in Crozier's laboratory had been Gregory Pincus, the experimental biologist who went on to develop the so-called birth control pill. Thereby, Skinner said, under Crozier's direction "Pincus worked on the control of biology, whereas I worked on the control of behavior." In reading Mach, Skinner was particularly struck by the idea that scientific concepts have adhering to them obscurantist traces of their earlier versions; the task of current practitioners, both believed, was to release them from the grasp of "metaphysical obscurities," as Mach had put it.

In writing his doctoral thesis, the young Skinner saw a way of applying the Machian point of view to the clarification of such concepts as the "reflex" of intact organisms, something he considered to be as basic in psychology as, say, mass is in physics. As Skinner recollected, he was "following a strictly Machian line, in which behavior was analyzed as a subject matter in its own right as a function of environmental variables *without reference to either mind or the nervous system.*" That was "the line that Jacques Loeb . . . had taken."[40] In this radically empiricist mode, the study of behavior reduced itself for Skinner, to start with, to the observation of the motion of the foot of a food-deprived rat, pressing down a small lever in an experimental box of standard size. Explanation was reduced to description, causation to the notion of function, and the chief goal was the correlation between observed events.

Skinner's doctoral thesis is still in the Harvard Archives (and differs in some details from the portions that were later published). There, Skinner lists his intellectual debts straightforwardly, beginning, "The reader will recognize a method of criticism first formulated in respect of scientific concepts by Ernst Mach"; he then draws attention to only five books: Mach's *Mechanics* and *Analysis of Sensations,* two books by Henri Poincaré, and Percy W. Bridgman's recently published *Logic of Modern Physics* (1927). Skinner's dissertation was the start of a career that continued without deviation along the same direction for over five decades; the compass had been fixed by his first contacts with Loeb's and Mach's books.[41]

Philipp Frank (1884–1966)

Skinner was probably the last scientist who could say he followed "a strictly Machian line," who could imagine having drunk directly from the pure source. As we penetrate further into the intellectual milieu in which American scholars existed about fifty years after James's solitary journey to meet Mach, we see an ever-increasing variety of intellectual debts and a multiplicity of interactions of like-minded scientists and philosophers. In order to indicate the rest of Mach's long-term, variously mutated effect on U.S. thinkers by concrete examples, we shall shortly focus on one contemporary scholar who is usually referred to as the dean of philosophy in America (and who, as it happened, was one of Skinner's fellow students at Harvard during the 1930s), W. V. Quine. But to set the stage for understanding his early development, we must first consider some professional and personal relationships that existed during the first decades of this century in central Europe.

At the center of that network we find the physicist and philosopher Philipp Frank, a man who by training, imagination, and personality seemed selected by fate to play a key role in the wider transmission, reformulation, and transmutation of Mach's ideas. Born in Vienna in 1884 and educated as one of Ludwig Boltzmann's last students, he came to know Mach closely; was one of the originators of the group that became the Vienna Circle and the movement of twentieth-century scientific empiricism; was called to the university in Prague—where Mach had been active for twenty-eight years and had left behind a loyal band of admirers—in 1912 as Einstein's successor; taught there in Mach's spirit for twenty-seven years; wrote some of the most sympathetic accounts of Mach's work and influence; remained one of the most active members of the Vienna Circle movement, organizer of some of its international meetings, indefatigable author, editor, and academic politician; and finally, with the great dispersal of European intellectuals in the 1930s, went to America, where he headed its successor movement in its various manifestations, including as president of the Institute for the Unity of Science. On his standing as a physicist, we have Einstein's testimony that he valued Frank so highly that he recommended no other successor when he left the University of Prague in 1912.[42] Of Frank's work in the philosophy of science, it has been said, correctly, that it "combines informal *logical* analyses of the sciences with a vivid awareness of the *psychological* and *social-cultural* factors operating in the selection of problems, and in the acceptance or rejection of hypotheses, which contribute to the shaping of styles of scientific theorizing. In a sense, this is a genuine sequel to the work of Ernst Mach."[43]

Much has been written about the Vienna Circle, its early debts to Mach, and the various later cross-currents and phases of the movement; lately there have also been signs of a considerable increase in interest in it by a new generation of scholars. At least a summary of its origins and fate will be needed here, with special attention to the role of Philipp Frank, a personal intermediary between Mach and his younger contemporaries, between the epistemologies of Mach and of those who succeeded him, and between the Europeans and the interested Americans.

Then a young Privatdozent in Vienna, Frank began in 1907 to meet regularly on Thursday nights in one of the old Viennese coffeehouses with a small group; it consisted of some students as well as Hans Hahn (later professor of mathematics at the University of Vienna) and the political economist and sociologist Otto Neurath (later organizer of adult education in social science for the city of Vienna). Others, such as the scientist Richard von Mises, joined them occasionally. Their long, informal discussions on current problems of philosophy and science, and particularly on the relation of reason and experience, resembled those in other early twentieth-century study circles of young intellectuals. The aim of these evening sessions, according to Frank, was to "bring about the closest possible *rapprochement* between philosophy and science," and also to avoid "the traditional ambiguity and obscurity of philosophy."[44]

One of the first books on which this group seems to have centered attention was Abel Rey's *La théorie de la physique chez les physiciens contemporains* (1907), which, with extensive commentary on William Rankine, Mach, Ostwald, and Poincaré, announced a crisis of contemporary physics due to the failure of the mechanistic view that had been central to nineteenth-century physics. According to Rey, there was now no ontological basis left for science—it was only a collection of empirical recipes. As a consequence, faith in science as such was undermined. Frank recalled that many readers thought it necessary to "return to the medieval ideas that may be characterized as the organismic conception of the world," with its religious implications. To Frank and his friends this prospect was intolerable. "In this critical situation our minds turned toward a solution that had been advanced about 25 years before by our local physicist and philosopher, Ernst Mach, namely that explanation is to be sought not by means of mechanistic or organismic conceptions, but only by giving a descriptive account of the phenomenon. In this way, Mach . . . saved the scientific worldpicture from going down along with the mechanistic picture."[45]

This group, like others, was not without its own criticism of Mach's views. With the advance of science it had become clear that Mach gave an insufficient role to logic and to mathematics (as will be further discussed in Chapter 2), and had underestimated the fruitfulness of the atomistic hypothesis. The group therefore decided to build on, and as necessary to recast, Mach's ideas, to bring them into conformity with the modern situation as well as with the writings of related authors such as Poincaré and Duhem.

What the group fully approved was the anti-metaphysical tendencies launched by Mach, which they felt to be not only a requirement for better logic but also of "great relevance for the social and cultural life."[46] They saw Mach's function as analogous to that of the eighteenth-century Enlightenment figures and spoke of reading Mach "drunk with soberness"—the phrase often applied to Voltaire. On the other hand, while they joined in Machian empiricism as a starting point, the group as a whole showed less interest in Mach's forays into the history of science, although Frank was and remained an exception in this respect, as Carnap later attested.[47]

Perhaps the first product of this discussion circle was Frank's public debut on the scene, his 1907 article on causality and experience, "Kausalgesetz und Erfahrung" (which he later expanded into the widely read book *Das Kausalgesetz und seine Grenzen*). It was clearly written under the influence of Mach and Poincaré, both of whom would have agreed with much in it. It quickly aroused comments from two very different quarters: V. I. Lenin, who savaged it, and Albert Einstein, who became a lifelong friend.[48]

It would be interesting to know precisely when Mach came to know Frank, for that contact—which "helped seal Frank's lifelong loyalty to most aspects of Mach's philosophy of science"—started a relationship of the greatest importance for the propagation of Machian ideas in Europe and the United States during the next six decades.[49] The case illustrates again—as for James, Loeb, Einstein, and many others—the power of Mach's ideas and personality in captivating rising young scientists.

The earliest known personal interaction between Mach and Frank came in 1910, the result of Mach's growing impatience in trying to understand Einstein's and Hermann Minkowski's work on relativity. Frank, recommended as an expert, visited Mach and accepted the request to prepare an introductory-level essay on the new theory. These details will take on some importance in the account in Chapter 2 of Mach's changing attitude to Einstein's physics. What most interests us now is that this first encounter set the tenor of the bonding between Mach and the man who, as will be shown, was to play such a large role in the transformation and transmigration of Machian teachings.[50]

A Harvest of Mach's Seeds

At the end of World War I, with the establishment of new democratic republics in central Europe and the general desire, at least among the younger generation, to bring their civilization into a new, modern phase, Mach's ideas continued to have a special attraction among intellectuals in Vienna even before the formation of the Vienna Circle around Moritz Schlick.[51]

Brought to Vienna in 1922, Schlick, originally a student of Max Planck but with a very different philosophical orientation, was indeed a worthy occupant of the philosophy of science chair at the university that had been held by Mach and for two years by Boltzmann. With Schlick as a new intellectual center in Vienna, the formerly open, Thursday-night conversations at the Café Central became well-structured, closed Thursday-night seminars in which Frank was a commuting member from Prague.[52] The meetings, involving highly competent individuals from a great variety of professions, occasionally allowed invited guests, such as the Americans to be noted shortly; this opportunity for making acquaintances would have long-range consequences. Another significant event was the arrival in 1926 of the former student of Gottlob Frege, Rudolf Carnap, brought to Vienna as Privatdozent in 1926 by Schlick. Carnap, who had been much influenced by Bertrand Russell and Alfred North Whitehead, published his seminal book, *Der logische Aufbau der Welt,* in 1928, and this, in Frank's words, was seen by "our Viennese group" as the long-hoped-for "integration of Mach and Poincaré."[53]

Because of Carnap's key position, together with Frank, in the later, American phase of the movement, it will be useful to glance here at his book. Its debt to Mach, together with Avenarius, Poincaré, Russell, and Whitehead, is indicated in its first pages. The "elementary concepts" of Carnap's system are immediate sense impressions and the relations of similarity and diversity between them. Frank tells us that the book also, to his and Carnap's own astonishment, reminded them strongly of William James's pragmatism—for example, "that the meaning of any statement is given by . . . what it means as a direction for human behavior"—and thus showed a promising affinity of their movement with "kindred spirits beyond the Atlantic in the United States."[54]

In his preface Carnap explained that a main impulse was to "banish metaphysics from philosophy, because its theses cannot be rationally justified"; and on the second page of the text, as if touching on another of Mach's main tasks and echoing the key notion of the *Aufruf* of 1911–1912, he declared his intention to be the construction of a system of concepts not only of natural science but of one total knowledge, a *Gesamtwissenschaft:* "Only if it becomes possible to build up a unified system of all concepts will it be possible to overcome the splintering of the *Gesamtwissenschaft* into separate part-sciences that stand, one next to the other, without relationship." In this way it would be possible to attain an "intersubjective, objective world . . . identical for all observers," and so make, as it were, an end run around supposedly essential differences between physics, biology, psychology, and so forth.[55] The desired unification was to encompass all fields of science and scholarship generally, and unity among them came to be looked for in terms of a commonality of concepts, of laws, of methods (including the unmasking of "meaningless" problems), and of the social community of researchers.

The movement was entering its most intense period. November 1928 saw the founding of the "Verein Ernst Mach" as the "official" forum of the Vienna Circle, with the stated goal to "further and propagate a scientific world view" and to achieve a unified body of knowledge, an *Einheitswissenschaft.* (A counterpart organization to the circle around Schlick had been formed in Berlin, renamed in 1931 "Gesellschaft für wissenschaftliche Philosophie.") The very name of the Verein demonstrated its intellectual debts, despite the differences mentioned earlier. The circle's devotion to logic and to the clarification of the meaning of propositions accentuated a range of additional influences on it, from Brentano to Russell and Ludwig Wittgenstein. How direct and strong the ties now were with Mach, and with which aspects of Mach, became ever more discussable. Frank was perhaps the most faithful and persistent follower, calling "the harvest of the seeds scattered by Mach . . . particularly rich and in the strictest accordance with his true intentions." As a shrewd observer from the other side of the Atlantic put it on noting the place now given to logic: "Mach, it is true, has recently been canonized and made the father of a new school of philosophy in Vienna. . . . But this Mach *redivivus* is the positivistic and not the pragmatistic Mach."[56]

But sorting out unambiguously the differences with Mach's original teachings is extremely difficult, for two intimately connected reasons. As Mach always insisted, he had no coherent, easily categorized, and time-stable "philosophy"; and that made it easy for different scholars to attach themselves to different fragments or versions of an evolving point of view. Also, the Vienna Circle and its sympathizers elsewhere did not form a well-synchronized group of mere disciples but, rather, included a great variety of energetic and imaginative individuals with different backgrounds. Thus Carnap and Schlick could argue in their letters as late as 1926 about what Machism really meant, as did Neurath in his letters to von Mises even in 1939, with Mach's ghost hovering over the discussants.

The most spectacular and widely noted act of the group was the publication of the manifesto *Wissenschaftliche Weltauffassung: Der*

Wiener Kreis (1929), under the authorship of Carnap, Hans Hahn, and Otto Neurath. This document spelled out the group's doctrines in persuasive and generally accessible terms, taking great care to avoid wrong associations and warning against the confusion between the emotional and informational functions of language. The slim booklet announced a fundamental aim of the Vienna Circle philosophy to be anti-metaphysical, unified science ("Einheitswissenschaft"), in which every symbol denotes "something 'real' if it is coherent with the total structure of facts of experience."[57] It thus addressed what Carl G. Hempel, who studied in Vienna in 1929–1930, identified as the group's common aim of removing the "strong intellectual discomfort" that philosophy, unlike science, "had had so little success with its efforts to solve certain central problems, above all in metaphysics."[58] But its larger ambition was to be a tocsin for modernism, going far beyond the natural sciences and logic. It asked for a world conception inspired by Enlightenment ideas that encompassed the modernization of all life, from economics to architecture, from the education of workers to the formation of the tools of thought of modern empiricism that are needed for the conduct of "private and public life," as well as "business and social life." The memorable concluding sentence of this manifesto was, "The scientific world conception serves life, and in turn is taken up by life."[59]

Not all members of the circle were of one mind politically; the spectrum ran from the agitated Neurath to the almost apolitical Schlick. But for the most part they shared a revulsion against the oppressive residue of medievalism in so much of Austrian intellectual and political life, longed instead for a *neue Sachlichkeit* (a new sober factuality), and agreed in their liberal, secular extraphilosophical aims just as they were all fairly unified on the central philosophical ones. Carnap's autobiographical statement on this point is important, both because it records the predominant spirit and because it overlapped with the point of view of a number of Americans who later sponsored the immigration of key members of the circle.

> I have not been active in party politics, but I was always interested in political principles and I have never shied away from

> professing my point of view. All of us in the Vienna Circle took a strong interest in the political events in our country, in Europe, and in the world. These problems were discussed privately, not in the Circle which was devoted to theoretical questions. I think that nearly all of us shared the following three views as a matter of course which hardly needed any discussion. The first is the view that man has no supernatural protectors or enemies and that therefore whatever can be done to improve life is the task of man himself. Second, we had the conviction that mankind is able to change the conditions of life in such a way that many of the sufferings of today may be avoided and that the external and the internal situation of life for the individual, the community, and finally for humanity will be essentially improved. The third is the view that all deliberate action presupposes knowledge of the world, that the scientific method is the best method of acquiring knowledge and that therefore science must be regarded as one of the most valuable instruments for the improvement of life. In Vienna we had no names for these views; if we look for a brief designation in American terminology for the combination of these three convictions, the best would seem to be "scientific humanism."[60]

The wide variety of programs launched by the group is therefore not surprising; it included, for example, public lectures in Vienna such as Carnap's "On God and Soul: Pseudo-Problems of Metaphysics and Theology," and one by Philipp Frank's brother, the distinguished architect Josef Frank, entitled "Modern World Conception and Modern Architecture." As Peter Galison has persuasively argued, the ambition of Carnap's *Aufbau* was to be a manifesto of modernism, and the whole work of the Vienna Circle lent itself to an attempt at a new integration of science, philosophy, art, architecture, and social values.[61] Thus the welcome that adherents of the Bauhaus later would experience in the United States was not unconnected to that extended to the Vienna Circle members. Similarly, Herbert Feigl was dispatched in 1929 "as the first 'emissary' of the Vienna Circle to Bauhaus Dessau" because it was considered to have a related, progressive ideology. And feeling that there "was a Zeitgeist thoroughly congenial to our Viennese position" waiting across the ocean, Feigl exported the main principles of the circle to the United States in a paper he wrote with Alfred E.

Blumberg—a publication that in its title, "Logical Positivism: A New Movement in European Philosophy," also provided the movement "its international trade name."[62]

The Vienna Circle and the related Berlin assembly had felt frustrated to be a "small number of dissident people hemmed in by the vast ocean of German school philosophy which was more or less a development of Kantian metaphysics."[63] By 1929 the internal morale, energy, and ambition of the group were so high that an intense search for larger public forums ensued—the operational indication that a movement had emerged. One step was the founding of the journal *Erkenntnis,* the main mouthpiece of the movement, edited by Carnap and Hans Reichenbach (later, in its American phase, continued as the *Journal for the Unity of Science,* published by the University of Chicago Press). Also, two series of books were launched, with Frank as coeditor: *Schriften zur wissenschaftlichen Weltauffassung* (10 vols.) and *Einheitswissenschaft* (7 vols.); to these were added, starting in 1938, the Library of Unified Science Series and the International Encyclopedia of Unified Science.

Yet another move, also with important consequences, was initiated by Frank. As the professor of theoretical physics at what was now called the "German University" at Prague, he was chairman of the local committee of the 1929 annual meeting of physicists and mathematicians from German-speaking Europe, scheduled for Prague, as well as chairman of the physicists' portion of the congress. Therefore he could not be refused his request to include a session, jointly sponsored by the Ernst-Mach-Verein of Vienna and the Society for Empirical Philosophy of Berlin, and advertised as the First Congress for the Theory of Knowledge of the Exact Sciences. It seemed an ideal moment to try to convert the German physicists. In his own lecture at the opening session Frank traced the rise of scientific empiricism from Mach, indicated its overlap with James's pragmatism, and ended with the challenge, "There are no boundaries between science and philosophy if only one formulates the task of physics in accordance with the doctrines of Ernst Mach."[64]

If Frank had hoped to introduce the new philosophy to the scientists and thereby shake their "sentimental ties to Kantianism" immediately, he did not succeed.[65] The majority of the audience was hostile. Yet for the movement the Prague meeting became the model for a whole series of congresses under different titles (e.g., International Congress for the Unity of Science) in different countries: in 1930 in Königsberg; 1934 in Paris (with papers read by two Americans, Ernest Nagel and Charles Morris); 1935 again in Paris; 1936 in Copenhagen, with Niels Bohr's involvement (see Figure 2); 1937 again in Paris; 1938 in Cambridge, England (with

Figure 2. A session at the Second International Congress for Unity of Science, Copenhagen, 21–26 June 1936, on the topic, "The Causality Problem." (Courtesy of Harvard University Archives, P. Frank file.) In the front row are Philipp Frank *(left)* and Niels Bohr *(right)*. Between and behind them are Harald Bohr and Georg von Hevesy; and, to Bohr's left, the educator Hannah Adler and the psychologist Edgar Rubin. Among others visible are Joergen Joergensen (standing), Otto Neurath (third from left in fourth row), Carl Hempel (behind him, toward the right), and Karl Popper (third from Hempel to the right). For the participation of Werner Heisenberg, Pascual Jordan, and other German scientists, see Dieter Hoffmann, "Zur Teilname deutscher Physiker an den Kopenhagener Physikerkonferenzen," *NTM: Zeitschrift für Geschichte der Naturwissenschaft, Technik und Medizin,* 25 (1988): 49–55.

papers by Max Black, V. Lenzen, and D. C. Williams from the United States); September 1939 in Cambridge, Massachusetts, and 1941 in Chicago.

There were two other important results of that first meeting. One was the establishment of the self-confident internationalism of the movement, which would have important benefits when foreign acquaintances were later called upon to help in transplanting scholars in search of a refuge. More immediately, in the aftermath of the congress, Frank succeeded in creating at the University of Prague a special professorship of the philosophy of science and in having Rudolf Carnap appointed to it. Carnap's arrival in 1931 fortified this outpost of empiricist philosophy on Mach's own home ground. "From 1931 on, we had in this way a new center of the 'scientific world conception' at the University of Prague."[66] As it turned out, they also had another magnet to attract American visitors.

With Germany still largely in the sway of Kantian idealism, the groups in Vienna and Prague now thought it all the more important to seek allies in Great Britain, to a certain extent in France, and above all in the United States, where, as noted, the ground had been prepared by the work of Peirce, James, and to some extent John Dewey and others; by the writings of more recent sympathizers there, such as Bridgman at Harvard and Morris at Chicago; and by the travels of Schlick and Feigl. There started a flow of visitors from America. Among those who came to Vienna and Prague to learn and discuss were Nagel, Morris, Dickinson S. Miller, and a young man named W. V. Quine.

One may well ask what was so special about the European group that it reached out to young intellectuals on the other side of the Atlantic. At least two forces were at work. One was the lack of major figures in America itself; Feigl, who had emigrated there in 1931, observed that while there were a few important philosophizing scientists (and even they were on the whole isolated and spread across the continent), "perhaps the only prominent American philosopher of science after C. S. Peirce" was Morris Raphael Cohen of City College in New York; and Cohen and A. C. Benjamin were "the only really distinct representatives and teachers of the philosophy of science" in the whole country.[67]

This lack would be filled within a decade, in large measure by immigrants from the Continental group and their students. The second factor that increased the attraction of Vienna, Prague, and associated centers for these young philosophers was the remarkable wealth of philosophers of science of various related schools in German-speaking Europe, particularly those of Austrian origin. (Just among the latter, Nyíri lists, in roughly chronological order, the following: Bernard Bolzano, Mach, Karl Menger, Boltzmann, Alois Höfler, Edmund Husserl, Wittgenstein, Hahn, Neurath, Feigl, Frank, Popper, Ludwig Fleck, von Mises, Michael Polanyi, and others.) It is still rather puzzling what produced this efflorescence despite the fact that the university in Vienna tended to marginalize these scholars; but their presence and perhaps their very marginalization created a hospitable ambience for foreign visitors.[68]

During the early 1930s, the movement's most confident period, the Prague branch continued to be led by Frank, who was now director of the Institute of Theoretical Physics. Frank and Carnap shared an office that once had been Einstein's. As another student of Frank, Peter Bergmann, later recalled, the institute was located on the top floor of the academic building at Viničná 3, "conveniently across the street from the psychiatric hospital . . . The patients would look at us, and we at the patients, often wondering who was 'in' and who was 'out.'"[69] But here a reminder is in order that the clouds of madness were now gathering over Europe, heralding a tragic ending for many intellectuals of the movement and of the whole inheritance of Mach's influence on thought on the Continent. In 1936 Moritz Schlick was killed by an enraged student on the steps of the University of Vienna. Two years earlier the protofascist government of the Austrian Chancellor Engelbert Dollfuss had dissolved the Ernst-Mach-Verein in a police action, charging that it had been politically engaged on the side of the Social Democrats.

W. V. Quine

Having set forth the necessary topography and time line, we can pick up the thread of the early development of W. V. Quine, now widely regarded as the central figure in the philosophy of the post-positivistic era, the synthesizer of the problems of language versus theory and science versus philosophy.[70] In Quine's philosophy, as he has pointed out, the central question is, "Given only the evidence of our senses, how do we arrive at a theory of the world?" Starting with his essays "Truth by Convention" (1935) and "Two Dogmas of Empiricism" (1951), he sought the answer in what has been called a holistic or naturalistic version of empiricism that can trace a line of descent from Auguste Comte, Mach, and the Vienna Circle. To put it very briefly, the so-called Duhem-Quine thesis declares that only the body of a theory as a whole can be properly subjected to empirical test, rather than each isolated hypothesis. As Quine wrote,

> A conflict with experience at the periphery occasions readjustments in the interior of the field . . . Having reevaluated one statement we must reevaluate some others, which may be statements logically connected with the first or may be statements of logical connections themselves. But the total field is so underdetermined by its boundary conditions, experience, that there is much lattitude of choice as to what statements to reevaluate in the light of any single contrary experience.[71]

In essence, Quine's position can be seen as a critique and a restructuring of the logical empiricism of the Vienna Circle, so to speak from the inside, and particularly, as Quine has acknowledged, as a result of his contact with Rudolf Carnap: "I, like many, have been influenced more by him than by any other philosopher." In his searching analysis, Dirk Koppelberg sees Quine as the synthesizer of Carnap's and Neurath's ideas—a "continuation and finalization" of the empiricism of the Vienna Circle.[72]

From his autobiographies, we know Quine's personal preparation. In his high school years he had read James's *Pragmatism* "compulsively" (i.e., uncritically), and at Oberlin College, where he was studying mathematics, he was exposed to the work of John B. Watson in a psychology course and discovered Russell. By 1930 he was at Harvard for graduate work, and there he met Herbert Feigl, who had come on a Rockefeller Foundation fellowship to study for nine months under Bridgman. Feigl recalled encountering the philosophers C. I. Lewis, Henry Sheffer, Susanne K. Langer, and Alfred North Whitehead; but he was "especially impressed" with Quine. One result was that when Quine won a traveling study fellowship for 1932–1933, he took Feigl's advice "to start the year at Vienna"; he was also urged to go there by a fellow student, John Cooley, "who had discovered Carnap's *Logische Aufbau der Welt.*"[73]

A road had been chosen. Quine arrived in Vienna in September 1932 for a five-month stay, and at once plunged into exciting waters—attending Schlick's lectures and, at Schlick's invitation, the weekly Vienna Circle evenings (the first talk he heard was Friedrich Waismann's report on Bridgman's *Logic of Modern Physics*). He met there members such as Kurt Gödel, Karl Menger, Hans Hahn, Olga Hahn-Neurath, Gustav Bergmann, and visitors such as Hans Reichenbach and A. J. Ayer. Quine also saw Schlick and his American wife socially, and even lectured on his doctoral thesis at one of the Vienna Circle meetings.

But he had missed Carnap, who had moved to Prague. Therefore Quine went on to Prague for six weeks in the winter of 1933, at Carnap's invitation. There he heard Philipp Frank lecture and "eagerly attended" Carnap's lecture course at the Physics Institute. No doubt attracted by Quine's qualities, Carnap opened up to the young man, allowing him in his seminar and giving him his articles and books, including the recent *Logische Aufbau.* This contact was for Quine "my most notable experience of being intellectually fired by a living teacher rather than by a book."[74] Quine takes credit for bringing news of Carnap's work back to Harvard by giving several lectures on it. When he was appointed instructor at Harvard—after spending three years (1933–1935) as a colleague of B. F. Skinner in Harvard's newly founded elite Society of Fellows—he taught what he called "a philosophy course along Carnap's lines."[75] A new torch had been lit.

When the increasing persecutions in the late 1930s brought more of the European intellectuals to the United States, Quine valued their reinforcing but ecumenical fellowship. He was secretary of the fifth International Congress for the Unity of Science, a summit meeting of scholars sympathetic to the movement, held at Harvard on 3–9 September 1939—just as the war broke out in Europe.

The congress was opened by a greeting by the President of the University, James B. Conant, and papers were presented by, or for, a group so distinguished and varied that it is appropriate to list the better known among them:

A. C. Benjamin	Janina Lindenbaum-Hosiasson
R. Carnap	R. B. Lindsay
A. Church	Hans Margenau
G. de Santillana	R. von Mises
H. Feigl	Charles Morris
P. Frank	Ernest Nagel
Kurt Goldstein	Otto Neurath
H. Gomperz	F. S. C. Northrop
K. Grelling	Paul Oppenheim
C. G. Hempel	Talcott Parsons
L. J. Henderson	W. V. Quine
Sidney Hook	Hans Reichenbach
Werner Jaeger	Louis Rougier
Joergen Joergensen	George Sarton
H. M. Kannen	S. S. Stevens
A. V. Karpov	Alfred Tarski
Felix Kaufmann	F. Waismann
Hans Kelsen	D. C. Williams
Susanne K. Langer	Robert S. Woodbury
Kurt Lewin	Edgar Zilsel

Some, including von Mises, had come to the United States as visitors to attend the congress; but now they would remain, increasing the presence and power of the movement as it developed in its American phase. Of this congress, Quine later wrote simply: "Basically this was the Vienna Circle, with accretions, in international exile."[76] If a date is needed to mark it, one may regard this as the moment when Mach's spirit at long last found a resting place in the New World.

The Vienna Circle in Exile

Between roughly 1940 and 1960, the movement for a scientific philosophy in the United States flourished, pushed forward especially by the influx of the arrivals from Europe. The demons against which Mach and his contemporaries had battled were long since chased out of science, and the basic role of empiricism and logic in modern philosophy seemed secure. The main direction of the movement brought over from Europe was now identified most often by the slogans "Unity of Science" and "Unified Science," versions of the old terms *Einheitswissenschaft* and *Gesamtauffassung* that had animated the manifestos of 1911–1912 and 1929 as well as Carnap's *Aufbau.* These concepts in turn had roots in the perception-based phenomenalistic monism of Mach that had so appealed even to Carus. (Indeed, the Unity of Science Movement, as it now wanted to be known, came to refer to itself as "Monism free from Metaphysics.") It was Mach more than anyone else who had promised the elimination of the boundaries between the separate sciences; in his inaugural lecture in Vienna in 1895 he had put it picturesquely: "As the blood in nourishing the body separates into countless capillaries, only to be collected again and to meet in the heart, so in the science of the future all the rills of knowledge will gather more and more into a common and undivided stream." Frank called Mach "the spiritual ancestor of the Unity of Science Movement" and urged the adoption of Mach's program as that of "our Unity of Science Movement, of our Congresses, and of our Encyclopedia."[77] It was, so to speak, another way of standing Hegel on his head: unification not through metaphysics but through the elimination of metaphysics.

Various instrumentalities dedicated explicitly to such activities had been developing for some years in Europe and were now ready to deploy in America. Otto Neurath, the originator of a grand International Encyclopedia of Unified Science, and until his death in 1945 its central organizing champion, had planned the project at least as early as 1920. The initial ambition for it was breathtaking: the archival collection of "Papers of the Unity of Science Movement" at the Joseph Regenstein Library of the University of Chicago indicates that two hundred encyclopedia volumes were planned, as well as ten supplemental volumes of a "Visual Thesaurus." The first collective "monograph," later to become the first of the nineteen "chapters" in the only two volumes of the encyclopedia that were actually published (under the title *Foundations of the*

Unity of Science), came out in 1938, containing essays by Bohr, Carnap, Dewey, Morris, Neurath, and Russell. The published volumes summarized the movement's status in the period between 1938 and the 1960s as clearly as the *Aufruf* and the Vienna Circle manifesto, respectively, had done in their time, and they also explicitly invoked the ancestral link, with Neurath claiming in the first chapter of Volume 1 of the Encyclopedia to be "continuing the work of Ernst Mach."[78]

A second instrument of the movement was the Institute for the Unity of Science, founded by Frank and run under his presidency from 1947 on lines similar to those of the Ernst-Mach-Verein. Conducted under the aegis of the American Academy of Arts and Sciences in Boston (where Frank was elected a Fellow in 1943) rather than any university, the Institute used the Academy's *Proceedings* for some of its publications. Those links were not accidental but were yet another example of the symbiosis between the Europeans' urge toward *Einheitswissenschaft* and similar American tendencies. As Frank explained later, the distinguished literary historian Howard Mumford Jones, on succeeding the astronomer Harlow Shapley as president of the Academy, had expressed the hope in his October 1944 inaugural address of overcoming "the fractation of knowledge" through the encouragement of the "pressures toward unity," for which the Academy, embracing members of all scholarly disciplines, seemed particularly well suited. A committee of the Academy to implement Jones's idea was soon calling for programs that would support the "synthesis of knowledge."[79] In founding the Institute, Frank and his colleagues provided one of the more visible responses to this call.[80]

The energy and persuasiveness of the leaders of the movement were enormous. One example of the perspicacious expansion of their hold on the attention of American scholars was a letter of 29 October 1950, conveyed by Morris on behalf of the Institute to Robert K. Merton at Columbia University. The Institute said that it planned to issue bibliographies on key fields of interest; therefore, the letter continued, "we wish very much that you would do one on the sociology of science." So years before that field had begun to draw general attention in academe, the Institute had targeted it, as well as the obvious person to undertake a bibliography.[81]

As important for the Institute's impact as its publications—including separately issued volumes by Frank and von Mises—were its regularly scheduled open meetings, usually held at the House of the Academy or in one of the universities' faculty clubs in the Boston area. These often resulted in vigorous and memorable discussions among attendees with varying degrees of allegiance to the movement; they included Henry Aiken, George D. Birkhoff, E. G. Boring, Bridgman, Karl Deutsch, Giorgio de Santillana, Frank, Roman Jakobson, Edwin C. Kemble, Gyorgy Kepes, Philippe Le Corbeiller, Wassily Leontief, Hans Margenau and Ernest Nagel (as visitors), Talcott Parsons, Harlow Shapley, B. F. Skinner, S. S. Stevens, Lazlo Tisza, Norbert Wiener, and Quine—who described one of these meetings in his autobiography, adding that they too appeared to him "in the way of a Vienna Circle in exile."[82] Just as in the earlier meetings in Europe, advanced students and young instructors with sympathy for the aims of the group were also encouraged to attend, perhaps in the hope that some of them would carry on the work in the future. Analogous meetings took place in Chicago, Los Angeles, Minneapolis, Berkeley, and Princeton.

An Ecological Niche for a Movement

There remains, finally, the need to return to a historical question with sociological overtones: What was it that made America, in roughly the middle third of this century, the most hospitable new home for the European descendants of nineteenth-century positivism? Although there were tragic victims of the persecution in Europe, and despite the well-known obstacles that scholars and scientists had to suffer initially, the number of members of the Vienna Circle and its associated groups in Prague, Berlin, Lvov, Warsaw, and elsewhere who eventually took up residence in the United States was substantial.[83]

The full answer is necessarily complex. In the first instance it includes, as previously indicated, the absence of predominant transcendental metaphysical philosophies and, on the contrary, the prior existence of analogous, native empiricist philosophical currents, of which the most recent was the "operationalism" ascribed to Bridgman and widely adopted by scientists after the 1927 publication of *The Logic of Modern Physics.*[84] But additional factors emerge from the records documenting the varied success of attempts by prominent refugees—such as Frank, von Mises, Reichenbach, Alfred Tarski, and Edgar Zilsel—to find academic positions.

To summarize, when favorable, the outcomes were in most instances the result of several interacting forces at work in the United States. At each of the universities that eventually provided a place for a refugee, there was at least one influential scholar who already knew of and respected the work of the candidate and undertook to labor on behalf of the cause. In this they were supported by recommendations from distinguished scholars at other American institutions. The university system, even during those difficult post-Depression years, was flexible enough, and some administrators sufficiently ingenious, occasionally to permit the creation of a variety of temporary, part-time, or externally funded posts that would often lead to more permanent arrangements.

A large share of the credit goes also to "unofficial," private organizations specifically created to help with advice and funds, exemplifying the American talent for self-organization—such as the Institute of International Education in New York, inspired by Alvin Johnson and others, and its Emergency Committee in Aid of Displaced Foreign Scholars, each run by a remarkable roster of concerned Americans. The Rockefeller Foundation, chiefly through Warren Weaver, was intensely active in providing support. The correspondence among the immigrants themselves also shows that on the whole they were realistic about the need to accommodate initially to much reduced circumstances, and that they formed a chain along which useful information about possible positions was passed. The disgust among Americans created by the persecutions visited on the European victims of fascism and by its program of cultural destruction provided additional energy; this had prompted Bridgman's famous "Manifesto" announcing the closing of his laboratory to visitors from totalitarian states.[85]

But by far the most important factor was the American sponsors' feeling of welcome for the special expertise and general point of view brought by Continental scholars. To make this concrete by an exemplary case, one need only study the archival materials connected with the placement of one of the chief energizers of the European movement in its new home, Philipp Frank; similar cases could be presented for many others.[86]

By sheer luck, Frank and his wife Hania were spared the fate of so many after the rape of Czechoslovakia in 1939. They had come on a visit to America in 1938, and Frank was making a lecture tour of twenty universities where scholars were interested in his discussion of logical empiricism. One of these was Harvard, where his chief contact was Bridgman, who had been there since his undergraduate days in 1900–1904. Bridgman had corresponded with Frank's colleague at Prague, Carnap, since 1934, after Carnap had sent him *The Unity of Science.*[87] Bridgman, while expressing some reservation about the limits-in-principle of logic as a tool in every conceivable situation, had responded with enthusiasm: "In general I have taken great satisfaction in the writings of the Viennese Circle, including many of your own, as being more nearly akin to my own views than nearly any other analytical writing with which I am acquainted, and this last book of yours is no exception." (They continued to correspond for years, and it is significant that one topic was the nature of "pencil-and-paper operations," which were giving Bridgman considerable intellectual discomfort.)

Frank first contacted Bridgman in a letter of 25 February 1938, noting that he had "always firmly agreed with your operationalist view" and expressing interest in including Harvard in his forthcoming lecture tour so as to have "the opportunity of discussing with you and your friends and students the rôle of operationalism in modern physics." Frank included some of his reprints and a copy of *Kausalgesetz* in a French translation.

Bridgman's reply of 30 March 1938—by which time Austria had welcomed the takeover by the Nazis, and Czechoslovakia was being threatened—is extremely revealing, for it casts light on similar situations at other universities that soon were to be offered refugee scholars. Bridgman wrote:

> I was glad to get your letter and to know of your projected visit to this country next fall . . . I read [the reprints and the book] with very great interest. It is naturally a source of gratification to me that we can agree on so many points . . .

> It will be a great pleasure to see you in Cambridge next fall and to talk things over with you. I am afraid you will not find Cambridge the center of activity with regard to the questions of interest to you which you apparently suppose. My work is done practically alone. I have no students [in philosophy of science] and have practically no contacts with members of the Department of Philosophy, and, in fact, most of them are not at all sympathetic with our point of view. The only young philosopher here whom I have particularly interested is Dr. Quine.

Bridgman's loneliness in philosophical matters—with which he struggled daily, even in his pioneering experiments in high-pressure physics—and Frank's evident excellence made Bridgman interested in bringing Frank into the physics department when that idea was raised—a scenario to be played out many times, at Harvard and elsewhere. Thus the theoretical physicist Edwin C. Kemble, Bridgman's former student and now a philosophically close colleague at Harvard, had read an article by Frank on philosophy of science and wrote him, on 4 January 1939: "I feel perhaps a closer bond of kinship with you with respect to these matters than with anyone else with whom I have talked."

Frank came to visit Harvard in December 1938 and lectured on "Philosophical Interpretations and Misinterpretations of Quantum Theory" under the joint auspices of the departments of physics and philosophy. The topic intersected with the continuing, real concerns of many scientists (including Kemble) and some philosophers (including D. C. Williams), concerns ranging from the status of unobservables to the nature of probability. This was just the kind of discussion that had been rare in the United States, while so prevalent on the Continent, for example in the interactions between Bohr, Einstein, Max Born, Erwin Schrödinger, Pauli, and Pascual Jordan with one another and with philosophers of science. Just as Mach had been found, in the words of H. A. Lorentz, to be essential "as a master and as a guide of their thoughts" through the thickets of late nineteenth- and early twentieth-century science, so Frank and his circle offered scholars in the United States guidance on more recent, persisting scientific and philosophical problems. Kemble diagnosed the needs in letters searching for funds with which to employ Frank, writing on 15 February 1939: "In recent years the borderland between physics and philosophy has come to be of increasing importance. It has become evident that clear ideas in physics cannot be had without the adoption of a correct, but far from naive philosophy. Theoretical physicists have become increasingly concerned with philosophy and philosophers have become increasingly influenced by the contributions of physicists." And he added: "Of all those with the training of the working physicist Frank is perhaps the most complete philosopher."

By spring 1939 Frank, now unable to return to Prague, was being considered for a one-year position at Harvard; he wrote to Bridgman from Chicago on 7 May that he looked forward to "the opportunity to collaborate with you and your department," to discuss "all the problems which belong to the so-called philosophical foundations of physics," to "help you to spread this spirit among the students of science," and to aid in the preparation of the "Unity of Science" congress at Harvard that Bridgman was to chair in September 1939. One recognizes echoes of the situation in which Frank had become so helpful to Ernst Mach himself, just thirty years earlier.

The relation between Bridgman and Frank was in fact quite symbiotic. Bridgman, with Kemble, spearheaded the presentation of the department's unanimous request (23 March 1939) that Einstein's successor at Prague be given the temporary position of unpaid research associate in physics and philosophy (1939–1940); they settled for this modest proposal because the Harvard administration was reluctant to add to the six refugees recently accommodated in various parts of the university. Bridgman also wrote on Frank's behalf to the Harvard University Press on 19 January 1940, urging publication of a translation of Frank's collected essays on the philosophical foundations of physics, which he regarded as a "most important project," "a valuable service . . . to the American public," for Frank impressed Bridgman "as perhaps the soundest" of the Europeans in that field. With the energetic help of Harlow Shapley, who ran a sort of underground railway to all parts of the United States to place European scholars fleeing fascism[88] and had contacts in the world of foundations, he also raised $2,000 to cover the initial year of Frank's stay (which Frank had to supplement with the advance payment by the A. A. Knopf Publishing Company for a projected biography of Einstein).

Once he was established more firmly in Cambridge—on a multi-year, half-time lectureship funded by monies to be raised by Shapley—Frank in turn used to the full his lively mind and the persuasive skills he had honed for decades, in the service of propagating scientific philosophy. In addition to teaching and writing, he presided over the numerous and various activities of the Institute, as noted, in which Bridgman and many of his colleagues took part. His effect on students and other colleagues was memorable; Shapley summed it up in a note to Frank dated May 1962: "You [were] my ghost thinker."

So by the fall term of 1940, the scene had changed dramatically from the lonely one depicted by Bridgman in his first letter to Frank two years earlier. Possibly the local readiness to give "modernism a chance" that had surfaced in the days of Josiah Royce, William James, and George Santayana was now interacting with the effect of the war in Europe, breaking down old habits of isolation. At any rate, when Feigl, at the University of Iowa since 1931, came back for a second Rockefeller Research Fellowship year at Harvard, he found the place transformed, with "fascinating regular discussions" in a group that included Frank, von Mises, Quine, Boring, Stevens, Bridgman, and I. A. Richards among the more active faculty members, as well as visitors such as Russell, Carnap, and Tarski. Like Quine, he observed: "There was a sort of revival of the Vienna Circle."[89] More important, the experience encouraged Feigl and others to launch collaborative teamwork in philosophical research, which had not been a familiar practice in the United States. Thereby, and through critical responses for and against aspects of Continental empiricism, philosophy in America changed vastly during the next decades. Few major universities would choose to remain without representation in the philosophy of science, in contrast to the paucity tolerated in the early 1930s that we noted before.

But starting in the 1950s, widely noted challenges to basic assumptions of logical empiricism arose out of the work of Quine and the late works of Wittgenstein, and also from two scholars with great influence in the history and philosophy of science, Alexandre Koyré and Norwood Russell Hanson. By that time, more than a decade of contacts with intellectual currents in America, as if by reaction, had begun to change the balance or direction of their movement. One can find an excellent example in Philipp Frank's own words as early as April 1950, when the Institute for the Unity of Science held its first national conference at the American Academy of Arts and Sciences, under Frank's chairmanship, and with the participation of President Conant of Harvard. In introducing the published proceedings of that conference, Frank wrote as follows:[90]

> The plan of this meeting was to discuss some issues which have been focal points in the approaches towards an integration of knowledge. During recent decades, substantial progress has been achieved by considering the sciences as formal systems and by analyzing them from the logical and semantical viewpoint . . . However, it has turned out more and more that these problems cannot be settled definitely on the basis of logical and semantical analysis. There remain always several possibilities for the choice of a formal system. Carnap contrasted recently in an excellent way the "internal" problems, which can be solved by logic and semantics, with the "external" problems. The latter ones put the question whether a certain formal system, as a whole, with the addition of a semantical interpretation, is useful for the orientation of man in the world of experience. Here we turn from the logical and semantical to the pragmatical viewpoint . . .
>
> What kind of argument do we call "pragmatic"? To get the answer we have to consider science as a human enterprise by which man tries to adapt himself to the external world. Then a "pragmatic" criterion means, exactly speaking, the introduction of psychological and sociological considerations into every science, even into physics and chemistry. It seems, therefore, that the sociology of science, the consideration of science as a human enterprise, has to be connected in a very tight way with every consideration which one may call logical or semantical.

By the time of Frank's death in 1966, the movement as such had run its course. The Institute and its activities essentially ceased, and the fierce focus of the movement's early years had given way to a dispersal and penetration in a variety of other versions of contemporary thought. To its critics, one might have applied Einstein's dictum that they were unaware how much they had imbibed of the belief system that they were now berating. Or one may perhaps say with Lewis A. Coser that "the Circle died of its members' success. Most of those one-time outsiders became insiders in America, and hence found it impossible to maintain their separateness."[91] But the continuing importance attached to logic, to analysis of language, to cross-disciplinarity, and to the other hallmarks of the Old World schools that shared in the post-Machian heritage is also a reminder of the role they played in helping the philosophy of science in the New World to rise to eminence.

What has also remained as one strain of contemporary philosophical thought is a commitment to the continued exploration of a science-based world view; it is characterized by a "critical attitude . . . [as] the basic condition for a sensible approach to the practical problems of our day" and by the hope that unsolved problems of philosophy now "can be formulated and treated with a precision and clarity formerly unknown." These phrases, by the philosopher Joergen Joergensen, appear on the final page of the last chapter of the final volume of the movement's encyclopedia, *Foundations of the Unity of Science.* And in his very last sentence, that author gives his correct judgment of the value of the movement at its end: "They have not created a new philosophical system, which, indeed, would have been contrary to their highest intentions, but they have paved the way for a new and fruitful manner of philosophizing."[92] The new generation may have expected more. But it is appropriate to remember that Ernst Mach had expressed his own aims for scientific philosophy throughout in just such terms.

Notes

1. Letters to the Nobel Committee on behalf of Mach from Eduard Suess, Ferdinand Braun, H. A. Lorentz, and Wilhelm Ostwald are printed in John T. Blackmore and Klaus Hentschel, eds., *Ernst Mach als Aussenseiter* (Vienna: Wilhelm Braunmüller, 1985). The letter of Lorentz (signed with W. H. Julius), 29 Jan. 1912, is on pp. 95–96, quoted on p. 96; Braun's letter, 24 Jan. 1911, is on p. 88. For the obituary see Albert Ein-

stein, "Ernst Mach," *Physikalische Zeitschrift,* 17 (1916): 101–104, on p. 102.

2. Albert Einstein, "Autobiographical Notes," in *Albert Einstein: Philosopher-Scientist,* ed. Paul A. Schilpp (Evanston, Ill.: Library of Living Philosophers, 1949), pp. 2–95, on p. 21; for the letter see Friedrich Herneck, "Die Beziehungen zwischen Einstein und Mach, dokumentarisch dargestellt," *Wissenschaftliche Zeitschrift der Friedrich-Schiller-Universität, Jena, Mathematisch-Naturwissenschaftliche Reihe,* 15 (1966): 1–14, on p. 6.

3. Moritz Schlick, "Ernst Mach," *Neue Freie Presse* (Vienna), Suppl., 12 June 1926, pp. 10–13, on p. 11. Unless otherwise noted, all translations are mine.

4. In recent years, there seems to have begun another cycle of interest in Mach and his influence, evidenced, e.g., by books such as Rudolf Haller and Friedrich Stadler, eds., *Ernst Mach—Werk und Wirkung* (Vienna: Hölder-Pichler-Tempsky, 1988); Dieter Hoffmann and Hubert Laitko, eds., *Ernst Mach: Studien und Dokumente zu Leben und Werk* (Berlin: Deutscher Verlag der Wissenschaften, 1991); Dieter Hoffmann, "Studien zu Leben und Werk von Ernst Mach" (doctoral diss., Humboldt-Universität zu Berlin, 1989; some of the material in this dissertation has been published in Hoffmann and Laitko, eds., *Studien und Dokumente*); Blackmore and Hentschel, eds., *Ernst Mach als Aussenseiter* (cit. n. 1); and Gereon Wolters, *Mach I, Mach II, Einstein und die Relativitätstheorie: Eine Fälschung und ihre Folgen* (Berlin, New York: Walter de Gruyter, 1987). As part of Chapter 2 we shall examine the last two of these books in some detail.

On Mach's claim not to be proposing a philosophy, see for example these two passages: "Therefore I have explicitly stated already that I am not a philosopher, but only a scientist. If nevertheless I am at times somewhat obstrusively counted amongst philosophers, the fault is not mine." "Above all, there is no Machian philosophy, but at best a scientific methodology and cognitive psychology, and both are provisional, imperfect attempts, like all scientific theories." Mach, *Knowledge and Error* (Dordrecht, Holland, and Boston: D. Reidel Publishing Company, 1976; from the 1905 edition and Mach's addition to it), pp. xxxii–xxxiii.

5. For example, outside science, Mach's influence has been documented on Arthur Schnitzler, Hermann Bahr, Richard Beer-Hoffmann, Hugo von Hofmannsthal (who heard Mach lecture at the University of Vienna), and Robert Musil. The effect of Mach's theory of epistemology on visual artists has been described by Joachim Thiele, "Zur Wirkungsgeschichte der Methodenlehre Ernst Machs," in *Symposium aus Anlass des 50. Todestages von Ernst Mach,* ed. W. F. Merzkirch (Freiburg

im Breisgau: Ernst-Mach-Institut, 1967), pp. 88–89. We know of the effect of Mach's thoughts on Walter Rathenau, F. von Hayek, Joseph A. Schumpeter, the young Wittgenstein, the young Heinrich Gomperz (later a classical philologist at Vienna), and many medical researchers, including the experimental pathologist Samuel von Basch. See Haller and Stadler, eds., *Werk und Wirkung,* particularly the essays by Stadler and Peter Mahr, for others among Mach's contemporaries for whom the impact of his ideas can be traced.

6. In an autobiography of 1910, cited by John T. Blackmore, *Ernst Mach: His Work, Life, and Influence* (Berkeley: University of California Press, 1972), p. 10. Mach also wrote: "America is the longing of my youth. I am greatly interested in the scholars and researchers there, among whom I have a number of friends." Letter of 20 Feb. 1899 to G. Stanley Hall.

7. For example, see the autobiographical sketch quoted in Hoffmann and Laitko, eds., *Studien und Dokumente* (cit. n. 4), p. 431.

8. Ernst Mach, "Die Leitgedanken meiner naturwissenschaftlichen Erkenntnislehre und ihre Aufnahme durch die Zeitgenossen," *Physikalische Zeitschrift,* 9 (1910): 599–606, on p. 604. For the police report see Blackmore, *Ernst Mach* (cit. n. 6), p. 83.

9. Pierre Duhem to Ernst Mach, 10 Aug. 1909, in Blackmore, *Ernst Mach,* p. 197.

10. One hundred thirty-six letters from the Open Court Publishing Company records of 1886–1930 were deposited by the Edward C. Hegeler Foundation and members of the Carus family in December 1968 at Southern Illinois University at Carbondale, Illinois; these can be found in the Morris Library Special Collections—Manuscripts. They include correspondence with Ludwig Mach. These holdings overlap with those at the Ernst-Mach-Institut at Freiburg im Breisgau.

I am grateful to the Curator of Manuscripts at the Morris Library, Sheila Ryan, for making copies of the correspondence with Ernst and Ludwig Mach available to me and for permission to quote excerpts. I also thank Blouke Carus, Paul Carus's grandson and president of the Open Court Publishing Company, for historical information.

11. Ernst Mach to Paul Carus, 22 Apr. 1895; all correspondence between Mach and Carus or other representatives of Open Court is in the Morris Library Special Collections cited in n. 10.

12. See, e.g., relevant passages in Ralph E. McCoy, ed., *Open Court: A Centennial Bibliography, 1887–1987* (La Salle, Ill.: Open Court, 1987), especially the "Historical Introduction" by Sherwood J. B. Sugden. To Sugden's bibliography on Carus and Open Court (p. 27), I would add

Joachim Thiele, "Paul Carus und Ernst Mach," *Isis,* 62 (1971): 208–219; and Thiele, *Wissenschaftliche Kommunikation: Die Korrespondenz Ernst Machs* (Kastellaun: A. Henn Verlag, 1978).

13. P. Carus, "Professor Mach's Philosophy," *Monist,* 16 (1906): 331–356, on p. 332; and Mach to Carus, 14 Apr. 1889. The work was difficult and slow, and the translation by Thomas J. McCormack (from the second German edition) appeared in 1893; but on 15 Feb. 1894 McCormack could assure Mach, "We have had some excellent reviews of the *Science of Mechanics.*"

14. Paul Carus, "Criticisms and Discussions," *Monist,* 16 (1906): 629. Mach's ideas, however, were not all that transparent to his commentators, and they also changed over time; thus in a letter to Carus, 7 June 1912, he expressed his doubts about monism as long as it had so many different meanings to different adherents.

15. Mach to Carus, 7 Jan. 1895; and Ludwig Mach to Carus, 28 Feb. 1913.

16. As Carus wrote Mach on 10 Nov. 1911, he and Edward Carl Hegeler (founder of the publishing company) had hoped early on to bring Ludwig to La Salle, to "see what he could do with American industrial work." On 27 Sept. 1889 Carus had written to Mach that it was a pity he could not accept the invitation of "Clark University in Wooster [*sic*], Massachusetts." On William Lang see Mach to Carus, 4 Sept. 1892.

17. Mach to Carus, 25 Dec. 1910.

18. Judith Ryan, "American Pragmatism, Viennese Psychology," *Raritan,* 8 (1989): 45–55, on p. 48; William James, *Essays, Comments, and Reviews* (Cambridge, Mass.: Harvard University Press, 1987), p. 297; and Ralph Barton Perry, *The Thought and Character of William James* (Boston: Little, Brown, 1936), vol. 2, p. 463.

19. William James to Mach, 27 June 1902, in Thiele, *Wissenschaftliche Kommunikation* (cit. n. 12), p. 172.

20. James to Mach, 19 Nov. 1902, ibid., pp. 173–174. Among evidences of selective disagreements see, e.g., several of Mach's letters to Anton Thomsen, in Blackmore and Hentschel, eds., *Ernst Mach als Aussenseiter* (cit. n. 1), pp. 86, 92, 111–113, in which he distances himself from aspects of James's publications; for James's own reservations see, e.g., William James, *Pragmatism* (Cambridge, Mass.: Harvard University Press, 1975), p. 34, where Mach is listed among authors who show that "human arbitrariness has driven divine necessity from scientific logic."

21. Various direct and indirect debts to Mach on the part of psychologists ranging from Edward B. Titchener to E. G. Boring have been noted; see, e.g., Blackmore, *Ernst Mach* (cit. n. 6); and Laurence D. Smith, *Be-*

haviorism and Logical Positivism: A Reassessment of the Alliance (Stanford, Calif.: Stanford University Press, 1986). One should certainly add here the psychophysicist S. S. Stevens.

22. Report on Schlick's talk, in *Erkenntnis,* 1 (1930): 75–76. Schlick added significantly: "One may regard John Dewey of Columbia University, New York, as a typical representative of American thought [in philosophy]. His philosophy . . . moves on the whole quite in the paths of empiricism in which Ernst Mach was a leader." Herbert Feigl, "The Wiener Kreis in America," in *The Intellectual Migration: Europe and America, 1930–1960,* ed. Donald Fleming and Bernard Bailyn (Cambridge, Mass.: Harvard University Press, 1969), pp. 630–673, on pp. 630, 661.

23. No letters from that decade have survived. But James had begun to express his admiration of Mach's work in print before they met; e.g., in 1880 he discussed Mach's "wonderfully original little work, '*Beiträge zur Analyse der Empfindungen*'": quoted in William James, *Principles of Psychology* (New York: Henry Holt, 1890), vol. 2, p. 50.

For the correspondence between Mach and James see Henry James, Jr., ed., *The Letters of William James* (Boston: Atlantic Monthly Press, 1920); and J. Thiele, "William James und Ernst Mach: Briefe aus den Jahren 1884–1905," *Philosophia Naturalis,* 9 (1966): 298–310, or its essentially identical reprinting in Thiele, *Wissenschaftliche Kommunikation* (cit. n. 12), pp. 168–176.

24. William James wrote to his wife on 2 Nov. 1882 that he had listened to Mach lecture on mechanics and found it "the most artistic lecture I ever heard." Their subsequent four-hour talk was "an unforgettable conversation. I don't think anyone ever gave me so strong an impression of pure intellectual genius. He apparently has read everything and thought about everything, and has an absolute simplicity of manner": in Thiele, *Wissenschaftliche Kommunikation,* p. 169. Similarly, after reading Ostwald's *Vorlesungen über Naturphilosophie,* James wrote to Hugo Münsterberg (23 July 1902), "I don't think I ever envied a man's mind so much as I have envied Ostwald's—unless it were Mach's": Perry, *Thought and Character of William James,* vol. 2, p. 288.

25. For example, Mach refers to James in fourteen places in *Analysis of Sensations,* often with laudatory comments, mostly concerning experimental results. For his part, James refers nine times to Mach in *Principles of Psychology,* again usually to experimental results, and in some cases at considerable length. Mach also appears in James's lecture notes for five philosophy courses, given at Harvard College over a period from 1879 to 1905; see William James, *Manuscript Lectures* (Cambridge, Mass.: Harvard University Press, 1988).

26. These letters from Mach to James are in William James Letters, Houghton Library, Harvard University, Cambridge, Massachusetts. They are printed in Thiele, *Wissenschaftliche Kommunikation* (cit. n. 12), pp. 168–176.

27. See the editorial comment in William James, *Some Problems of Philosophy* (Cambridge, Mass.: Harvard University Press, 1979), notes starting on p. 121. Mach's books surviving in that collection are *Grundlinien der Lehre von den Bewegungsempfindungen* (Leipzig: Wilhelm Engelmann, 1875), *Die Mechanik* (Leipzig: F. A. Brockhaus, 1883), *Analyse der Empfindungen,* 4th ed. (Jena: Gustav Fischer, 1909), *Populärwissenschaftliche Vorlesungen* (Leipzig: Barth, 1903), and *Erkenntnis und Irrtum* (Leipzig: Barth, 1905).

28. Attention has been drawn to three of the marginalia in the introduction by Erwin N. Hiebert to the reprinting of Ernst Mach's *Knowledge and Error* (Boston: Reidel, 1976).

29. James to Mach, 9 Aug. 1905, in Thiele, *Wissenschaftliche Kommunikation* (cit. n. 12), p. 175; and Ryan, "American Pragmatism" (cit. n. 18), pp. 52–53.

30. E.g., Blackmore, *Ernst Mach* (cit. n. 6); Hiebert, introduction to *Knowledge and Error* (cit. n. 28); Perry, *Thought and Character of William James* (cit. n. 24), vol. 2; Ryan, "American Pragmatism"; Susan Haack, "Pragmatism and Ontology: Peirce and James," *Revue Internationale de Philosophie,* 31 (1977): 377–400; Peter T. Manicas, "Pragmatic Philosophy of Science and the Charge of Scientism," *Transactions of the Charles S. Peirce Society,* 24 (1988): 179–222; Gerald E. Myers, *William James: His Life and Thought* (New Haven, Conn.: Yale University Press, 1986); and Hilary Putnam and Ruth Anna Putnam, "William James's Ideas," *Raritan,* 8 (1989): 27–44.

31. All three have been studied most recently in depth by Laurence D. Smith in *Behaviorism and Logical Positivism* (cit. n. 21), on which I shall rely to some extent.

32. For a good account of Loeb's life and work see Philip J. Pauly, *Jacques Loeb and the Engineering Ideal in Biology* (New York: Oxford University Press, 1987). An excellent, concise introduction to Loeb's thought has been provided by Donald Fleming in his introduction to the reprint edition of Loeb's *Mechanistic Conception of Life* (Cambridge, Mass.: Harvard University Press, 1964), pp. vii–xli.

33. Pauly, *Loeb and the Engineering Ideal,* p. 42. Writing to Einstein (31 Jan. 1924), Loeb recalled that "for many years Ernst Mach provided [for Loeb] the wonderful service" of banishing thoughts of discouragement or depression.

34. Pauly, *Jacques Loeb and the Engineering Ideal in Biology*, p. 5.

35. The main text was reprinted, although with an incomplete list of signers, in *Physikalische Zeitschrift*, 13 (1912): 735. Among its rare mention in secondary sources is a reference to its existence in Friedrich Herneck, "Albert Einstein und der philosophische Materialismus," *Forschung und Fortschritte*, 32 (1958): 204–208, on p. 206. I thank the Deutsche Akadamie der Wissenschaften zu Berlin for a copy of the original manifesto, from its Wilhelm-Ostwald-Archiv.

36. Headquartered in Berlin, both the Gesellschaft and the *Zeitschrift*, led by Petzoldt, lasted until 1915, the former was revived in 1927 as the Internationale Gesellschaft für empirische Philosophie, of which Petzoldt and Hans Reichenbach were members. See Blackmore and Hentschel, eds., *Ernst Mach als Aussenseiter* (cit. n. 1), p. 107.

37. A related document, "Gründe für die Bildung einer Gesellschaft für positivistische Philosophie," was reprinted in *Isis*, 1 (1913): 107–110, and in the *Journal of Philosophy, Psychology, and Scientific Methods*, 9 (1912): 419–420.

A striking omission in both documents is that of Ostwald. The copy in his archives has his annotation, "abgelehnt [refused]." An indication of how this affected Mach and Petzoldt, and how deeply they were involved in the *Aufruf*, emerges from a paragraph in one of Petzoldt's letters to Mach, dated 9 Jan. 1912. There he lauds Mach for having refused Ostwald's invitation to be honorary president of the Monistenbund and indicates that this was fair revenge: "Ostwald should now regret that he did not sign our *Aufruf*." The letter is quoted in Blackmore and Hentschel, eds., *Ernst Mach als Aussenseiter*, p. 100. For evidence that additional adherents to the *Aufruf* were expected, including Federigo Enriques, Poincaré, and Duhem, see *Revue Philosophique*, 76 (1913): 558–559; and Klaus Hentschel, *Die Korrespondenz Petzoldt-Reichenbach* (Berlin: Sigma, 1990), pp. 16–24.

38. B. F. Skinner, *The Shaping of a Behaviorist* (New York: Knopf, 1979); and Smith, *Behaviorism and Logical Positivism* (cit. n. 21), p. 277.

39. B. F. Skinner, "The Concept of the Reflex in the Description of Behavior," *Journal of General Psychology*, 5 (Oct. 1931): 427–457.

40. B. F. Skinner, review of Smith's *Behaviorism and Logical Positivism*, in *Journal of the History of the Behavioral Sciences*, 23 (1987): 204–209, on p. 209 (emphasis in original.)

41. "My debt was to the empiricism of Ernst Mach. If logical positivism can be said to have begun with the first issue of *Erkenntnis* [1929], I was far enough along in my own career to become a charter subscriber, as I was to its American equivalent, *Philosophy of Science*": ibid., p. 208.

42. For writings on Mach see e.g., his essays of 1917 and 1938, reprinted in Philipp Frank, *Modern Science and Its Philosophy* (Cambridge, Mass.: Harvard University Press, 1949), chaps. 2 and 3. For brief essays on Frank by eleven colleagues and a selected bibliography of his writings on the philosophy of science, see Robert S. Cohen and Marx W. Wartofsky, eds., *Proceedings of the Boston Colloquium for the Philosophy of Science, 1962–1964,* Boston Studies in the Philosophy of Science, 2 (New York: Humanities Press, 1965), pp. ix–xxxiv; see also the entry for Frank in the *Dictionary of Scientific Biography.* Einstein's evaluation appears in his handwritten draft (probably of 1937) of a recommendation for Frank, in Albert Einstein Archive, Jewish National and University Library, Department of Manuscripts and Archives, Jerusalem, doc. 11-087.

43. H. Feigl, "Some Major Issues and Developments in the Philosophy of Science of Logical Empiricism," in Feigl and Michael Scriven, eds., *Minnesota Studies in the Philosophy of Science,* vol. 1 (Minneapolis: University of Minnesota Press, 1976), p. 4.

44. Frank, *Modern Science* (cit. n. 42), p. 1.

45. Ibid., pp. 3, 6.

46. Ibid., p. 34. By "metaphysical" the circle members meant in-principle unverifiable and unfalsifiable.

47. "[Philipp Frank] was familiar with the history of science and much interested in the sociology of scientific activity, for which he collected comprehensive materials from history. Both because of his historical interest and his sound common sense, he was often wary of any proposed thesis that seemed to him overly radical, or of any point of view that seemed too formalistic. Thus, in a way similar to Neurath, he often brought the abstract discussion among the logicians back to the consideration of concrete situations": Rudolf Carnap, "Intellectual Autobiography," in *The Philosophy of Rudolf Carnap,* ed. Paul A. Schilpp (La Salle, Ill.: Open Court, 1963), p. 32.

48. Philipp Frank, "Kausalgesetz und Erfahrung," *Annalen der Naturphilosophie,* 6 (1907): 445–450; see also "Mechanismus oder Vitalismus?" ibid., 7 (1908): 393–409. Mach may well have read these, and also read or heard of Frank's public lecture of 4 Dec. 1909 at the Physikalische Gesellschaft at the University of Vienna on the topic "Does Absolute Motion Exist?" This piece was later published: Philipp Frank, "Gibt es eine absolute Bewegung?" in *Wissenschaftliche Beilage zum dreiundzwanzigsten Jahresbericht (1910) der Philosophischen Gesellschaft an der Universität zu Wien* (Leipzig: Johann Ambrosius Barth, 1911), pp. 1–19. In it, too, Frank explained, extended, and defended Mach's ideas.

Lenin commented on Frank in Chapter 3 of *Materialism and Empirio-*

Criticism (1909), as part of an attack concentrated mainly on Mach and Alexander Bogdanov. He dismissed Frank as a Kantian idealist. This attack might have been most uncomfortable, but happily Frank did not find out about it until the 1920s. Later, as he told me, Lenin's comment became useful to him in a completely unexpected way. While Frank was teaching at Harvard University, he was also doing consulting work for the U.S. Navy. Either in this connection, or as a result of the general anti-communist hysteria in the United States during the McCarthyite days after the war, Frank one day received a visit at his home from two FBI men. They had come to investigate his background and orientation, which seemed to them to have been suspiciously on the liberal side. Frank, no doubt with his usual quizzical smile, inquired whether they thought he might be a spy for the Russians, and to answer his own question, he went to his bookcase, fished out the copy of Lenin's book, and opened it to the passage where Lenin attacked him personally. As Frank ended this story, the two FBI men practically saluted him, and left speedily and satisfied.

In his first contact, Einstein made the objection that the simplicity of terminology in the law of causality, and therefore the "simplicity of nature," are *not* reducible to conventions. Frank learned the exchange that "logic needs a drop of pragmatic oil": Frank, *Modern Science,* p. 11.

49. Blackmore, *Ernst Mach* (cit. n. 6), p. 183.

50. See Philipp Frank, *Einstein: His Life and Times* (New York: Alfred A. Knopf, 1947) for Frank's personal comments on Mach.

51. A good witness to this was Friedrich von Hayek, who studied in Vienna in 1918–1921. He reported that his circle "sought arguments against metaphysics, which we found in Mach"; see W. F. Merzkirch, ed., *Symposium* (cit. n. 5), p. 42.

52. There are various accounts of the names of the members of the Vienna Circle at its height. Putting together the overlapping lists given by Otto Neurath, *Empiricism and Sociology,* ed. Marie Neurath and Robert S. Cohen (Dordrecht: Reidel, 1973), pp. 318ff., and Victor Kraft, *Der Wiener Kreis: Der Ursprung des Neopositivismus* (Vienna: Springer Verlag, 1950), pp. 3–4, one arrives at eighteen core members and nine strong sympathizers; but if one also adds active collaborators who published in major Vienna Circle programs, one would have to include such figures as Richard von Mises, then in Berlin. One estimate is that over a third of the total group eventually came to the United States.

Additional names of foreign visitors are given in Herbert Feigl, "Logical Empiricism," in *Twentieth Century Philosophy,* ed. Dagobert D. Runes (New York: Philosophical Library, 1943), p. 406.

53. Frank, *Modern Science* (cit. n. 42), p. 33. Similarly, Feigl wrote,

Carnap's *Aufbau* "seemed indeed the fulfillment of the original intentions of Mach's positivism, as well as a brilliant application of the tools of modern logic to some of the perennial issues of epistemology": "Wiener Kreis in America" (cit. n. 22), p. 635. The English translation, *The Logical Structure of the World,* was published in 1969 by the University of California Press. In observations such as Frank's and Feigl's, the correct implication of a somewhat indirect link between Mach and Vienna Circle positivism, which I have stressed throughout, differs from exaggerated claims of direct connections, such as Michael Polanyi's assertion that Mach's *Mechanik* of 1883 "founded the Vienna school of positivism" in *Personal Knowledge* (Chicago, Ill.: University of Chicago Press, 1958), p. 9. The most one can claim is that for the logical positivists Mach was a "model for philosophizing," as Klaus Hentschel puts it in his examination of Mach and his circle in *Interpretationen und Fehlinterpretationen der speziellen und der allgemeinen Relativitätstheorie* (Boston, Basel, Berlin: Birkhäuser Verlag, 1991), p. 368.

54. Frank, *Modern Science,* p. 33.

55. Rudolf Carnap, *Der logische Aufbau der Welt,* 1st ed. (Hamburg: Felix Meiner Verlag, 1928), p. xix, preface of May 1928, pp. 2–3. The hope expressed in the last phrase for an invariant description of the world, regardless of the observer, and for the removal of barriers between the specialty fields connects directly with Mach's views (as in n. 3) and also suggestively with Einstein's research program.

56. Frank, *Modern Science,* p. 89 (the essay was written in 1938); and Perry, *Thought and Character of William James* (cit. n. 24), vol. 2, p. 580. For a recent survey of the activities of the Verein, see Friedrich Stadler, "The 'Verein Ernst Mach': What Was It Really?" in John Blackmore, ed., *Ernst Mach—A Deeper Look: Documents and New Perspectives* (Dordrecht, Boston, London: Kluwer Academic Publishers, 1992), pp. 363–377.

57. Rudolf Carnap, Hans Hahn, and Otto Neurath, *Wissenschaftliche Weltauffassung: Der Wiener Kreis* (Vienna: Artur Wolf Verlag, 1929), pp. 15, 18 (my translation); for an English translation see Otto Neurath, *Empiricism and Sociology* (cit. n. 52). Frank explained that the word *Weltauffassung* was chosen to avoid the metaphysically charged and Germanic word *Weltanschauung* and that the subtitle "Der Wiener Kreis" was added at Neurath's suggestion to make the title "less dry" by evoking Vienna waltzes, the Vienna Wood, "and other things on the pleasant side of life": *Modern Science,* p. 38.

58. Carl G. Hempel, "Der Wiener Kreis: Eine persönliche Perspektive," in *Wittgenstein, der Wiener Kreis und der kritische Rationalismus,* ed. Hal Berghel, Adolf Hübner, and Eckehart Kohler (Vienna: Hölder-

Pichler-Tempsky, 1979), pp. 21–26, on p. 21. Among brief recent evaluations of the history of logical positivism, perhaps the most useful ones for students are in R. C. Olby, G. N. Cantor, J. R. R. Christie, and M. J. Hodge, eds., *Companion to the History of Modern Science* (London, New York: Routledge, 1990), chap. 54; Smith, *Behaviorism and Logical Positivism* (cit. n. 21), chap. 2; Robert N. Proctor, *Value-Free Science?* (Cambridge, Mass.: Harvard University Press, 1991), chap. 12; and Peter Achinstein and Stephen F. Barker, eds., *The Legacy of Logical Positivism* (Baltimore: Johns Hopkins Press, 1969).

59. Carnap, Hahn, and Neurath, *Wiener Kreis* (cit. n. 57), p. 30. Similarly, the fundamental aim of Richard von Mises's *Kleines Lehrbuch des Positivismus* (1939; Frankfurt am Main: Suhrkamp Verlag, 1990) was the renovation of culture in all its aspects.

60. Rudolf Carnap, "Intellectual Autobiography" (cit. n. 47), 82–83. For a debate about the role of politics in the Vienna Circle see the essays by Barry Smith and Gerhard Zecha in *The Vienna Circle and Lvov-Warsaw School,* ed. Klemens Szaniawski (Dordrecht: Kluwer, 1989).

61. These lectures were announced in *Erkenntnis,* 1 (1930–1931): 174. Peter Galison, "History, Philosophy, and the Central Metaphor," *Science in Context,* 2 (1988): 182–198; and Galison, "Aufbau / Bauhaus: Logical Positivism and Architectural Modernism," *Critical Inquiry,* 16 (1990): 709–752.

62. Feigl, "Wiener Kreis in America" (cit. n. 22), pp. 637, 645; Feigl and A. E. Blumberg, "Logical Positivism: A New Movement in European Philosophy," *Journal of Philosophy,* 28 (1931): 281–297 (Blumberg was one of the young Americans Feigl persuaded to come to study in Vienna); and Frank, *Modern Science* (cit. n. 42), p. 38. The group's self-identifying term *logical positivism* gave way, from about 1936, to *logical empiricism* or *scientific empiricism* for reasons sketched, e.g., in Feigl, "Wiener Kreis in America"; and Joergen Joergensen, "The Development of Logical Empiricism," in *Foundations of the Unity of Science: Toward an International Encyclopedia of Unified Science,* ed. Otto Neurath, Rudolf Carnap, and Charles Morris, vol. 2 (Chicago: University of Chicago Press, 1970), 845–936.

63. Frank, *Modern Science,* p. 47. For a debate why logical empiricism developed more easily in Austria than in Germany see Otto Neurath, "Le développement du cercle de Vienne et l'avenir de l'empirisme logique," *Actualités Scientifiques et Industrielles,* no. 290 (Paris: Hermann & Cie, 1936); Neurath, *Gesammelte philosophische und methodologische Schriften,* ed. Rudolf Haller and H. Rutte, 2 vols. (Vienna: Hölder-Pichler-Tempsky, 1981); Smith's and Zecha's articles in *Vienna Circle,* ed. Szaniawski (cit. n. 60); Haller, "Wittgenstein: An Austrian Enigma," in *Austrian Phi-*

losophy Studies and Texts, ed. J. C. Nyíri (Munich: Philosophia Verlag, 1981), pp. 91–112; Friedrich Stadler, *Vom Positivismus zur wissenschaftlichen Weltauffassung* (Vienna: Löcker Verlag, 1982); Carl G. Hempel and F. Stadler in *Wittgenstein,* ed. Berghel, Hübner, and Kohler (cit. n. 58); and several essays in J. C. Nyíri, ed., *Von Bolzano zu Wittgenstein* (Vienna: Hölder-Pichler-Tempsky, 1986). The contrast between philosophy as taught in the 1930s in Germany and the United States emerges graphically by comparing Sidney Hook, "A Personal Impression of Contemporary German Philosophy," *Journal of Philosophy,* 27 (1930): 141–160, and Charles W. Morris, "Aspects of Recent American Scientific Philosophy," *Erkenntnis,* 5 (1935): 142–150.

64. Philipp Frank, "Was bedeuten die gegenwärtigen physikalischen Theorien für die allgemeine Erkenntnislehre?" *Erkenntnis,* 1 (1930–1931): 126–157, on p. 157.

65. Frank, *Modern Science* (cit. n. 42), p. 40.

66. Ibid., p. 45.

67. Feigl, "Wiener Kreis in America" (cit. n. 22), p. 660. See also Daniel J. Wilson, "Science and the Crisis of Confidence in American Philosophy," *Transactions of the Charles S. Peirce Society* 23 (1987): 235–262.

68. J. C. Nyíri, "The Austrian Element in the Philosophy of Science," in *Bolzano zu Wittgenstein,* ed. Nyíri (cit. n. 63), pp. 141–146, on p. 142. The marginalization and mistreatment of these philosophers has been well documented in Friedrich Stadler, "Aspects of the Social Background and Position of the Vienna Circle at the University of Vienna," in *Rediscovering the Forgotten Vienna Circle,* ed. T. E. Vebel (Dordrecht: Kluwer, 1991).

69. Peter Bergmann, "Homage to Professor Philipp G. Frank," in *Proceedings of the Boston Colloquium* (cit. n. 42), pp. ix–x.

70. See Dirk Koppelberg, *Die Aufhebung der analytischen Philosophie: Quine als Synthese von Carnap und Neurath* (Frankfurt am Main: Suhrkamp Verlag, 1987). I shall base some of my points on Koppelberg's book and on Quine's own writings, including his autobiographical accounts, in *The Philosophy of W. V. Quine,* ed. L. E. Hahn and Paul A. Schilpp, Library of Living Philosophers, 18 (La Salle, Ill.: Open Court, 1986); W. V. Quine, *The Time of My Life: An Autobiography* (Cambridge, Mass.: MIT Press, 1985); and Richard Creath, ed., *Dear Carnap, Dear Van: The Quine-Carnap Correspondence and Related Work* (Berkeley: University of California Press, 1990). I also thank Professor Quine for comments on an early draft. See also his "Comment on Koppelberg," in *Perspectives on Quine,* ed. William Barrett and Roger F. Gibson (Oxford: Basil Blackwell, 1990), p. 212.

71. W. V. Quine, *The Roots of Reference* (La Salle, Ill.: Open Court, 1974), p. 1; and Quine, *From a Logical Point of View* (New York: Harper Torchbooks, 1963), p. 42 (in an essay written in 1953).

72. W. V. Quine, "Carnap's Positivistic Travail," *Fundamenta Scientiae,* 5 (1984): 325–334, on p. 333; and Koppelberg, *Die Aufhebung* (cit. n. 70), p. 20. On the other hand Quine, in "Comment on Koppelberg" (cit. n. 70), notes that he arrived at positions similar to Neurath's without traceable influence and, surprisingly, that he did not get from Duhem the holism associated with him but was alerted to Duhem only after the publication of his crucial 1951 essay "Two Dogmas"—"by both Hempel and Philipp Frank."

73. Hahn and Schilpp, eds., *Philosophy of Quine* (cit. n. 70), p. 6; Feigl, "Wiener Kreis in America" (cit. n. 22), p. 647; and Quine, *Time of My Life* (cit. n. 70), p. 86.

74. Quine, *Time of My Life,* p. 98. In his "Intellectual Autobiography" (cit. n. 47), p. 34, Carnap notes that Morris and Quine came to Prague: "Both were strongly attracted by our way of philosophizing and later helped to make it known in America."

75. One result of Quine's introduction of Carnap's work was Carnap's visiting year at Harvard in 1936 and the honorary degree awarded him at Harvard's Tercentenary Celebration that year. Charles Morris, who had spent the summer of 1934 in Prague, arranged for Carnap's professorship at the University of Chicago from 1936 and also helped to find positions in the United States for Carl Hempel and Hans Reichenbach; see Feigl, "Wiener Kreis in America" (cit. n. 22), p. 648. For the quotation see Hahn and Schilpp, eds., *Philosophy of Quine,* p. 16. The "Lectures on Carnap" (given 8–22 Nov. 1934) have been published in Creath, ed., *Dear Carnap, Dear Van* (cit. n. 70). The letters between Carnap and Quine show the symbiotic relation at work, while Carnap's "Intellectual Autobiography" tends to accentuate their eventual differences.

76. Hahn and Schilpp, eds., *Philosophy of Quine,* p. 19. The impressive list of papers presented at the September 1939 conference is given in the *Journal of Unified Sciences* (previously *Erkenntnis*), 8 (1939–1940): 369–371.

77. Ernst Mach, *Popular Scientific Lectures,* 5th ed. (La Salle, Ill.: Open Court, 1943), p. 261; and Frank, *Modern Science* (cit. n. 42), p. 89. For a good, brief discussion of the various meanings of "unity of science" current in the 1930s see Herbert Feigl, "Unity of Science and Unitary Science," *Journal of Unified Sciences (Erkenntnis),* 9 (1939–1940): 27–30.

78. Otto Neurath, "Unified Science as Encyclopedic Integration," in *Foundations of the Unity of Science,* ed. Neurath, Carnap, and Morris (cit.

n. 62), vol. 1 (Chicago: University of Chicago Press, 1955), pp. 1–27, on p. 14.

79. Philipp Frank, introductory remarks to the issue on "Contributions to the Analysis and Synthesis of Knowledge," *Proceedings of the American Academy of Arts and Sciences,* 80 (1951): 5–8, on p. 6. President James B. Conant of Harvard added his welcome to this effort (pp. 9–13) and linked the "quest for unity in science" with his plans for the "general education" program in science for undergraduates.

80. The Institute's charter of 31 July 1947, published in *Synthese,* 6 (1947): 158–159, specified: "The purposes for which the corporation is formed are to encourage the integration of knowledge by scientific methods, to conduct research in the psychological and sociological backgrounds of science, to compile bibliographies and publish abstracts and other forms of literature with respect to the integration of scientific knowledge, to support the International Movement for the Unity of Science, and to serve as a center for the continuation of the publications of the Unity of Science Movement." The Institute's background and purpose are discussed in detail by Frank in the pages that follow (ibid., pp. 160–167).

81. By March 1951 the bibliography had been prepared by Merton in collaboration with his former student Bernard Barber; it was published by Frank in the May 1952 issue of volume 80 of the *Proceedings* of the Academy. I thank Professor Merton for sharing copies of the correspondence.

82. Quine, *Time of My Life* (cit. n. 70), p. 219. I thank Professor P. R. Masani, now of the University of Pittsburgh, for supplementing my own memory and records of these meetings. As a graduate student he acted for a time as secretary of the group under Frank.

83. For a sober assessment of the experience of immigrant scholars see Paul K. Hoch, "The Reception of Central European Refugee Physicists of the 1930s: USSR, UK, USA," *Annals of Science,* 40 (1983): 217–246; also the essays by Nathan Reingold and by P. Thomas Carroll in Jarrell C. Jackman and Carla M. Bordon, eds., *The Muses Flee Hitler: Cultural Transfer and Adaptation, 1930–1945* (Washington, D.C.: Smithsonian Institution Press, 1983). Also relevant are Norman Bentwich, *The Rescue and Achievement of Refugee Scholars* (The Hague: Martinus Nijhoff, 1953); Stephen Duggan and Betty Drury, *The Rescue of Science and Learning: The Story of the Emergency Committee in Aid of Displaced Foreign Scholars* (New York: Macmillan Co., 1948); Robin E. Rider, "Alarm and Opportunity: Emigration of Mathematicians and Physicists to Britain and the United States, 1933–1945," *Historical Studies in the Physical Sciences,* 15 (1984): 107–170; and Christian Thiel, "Folgen der Emigration deutscher und

österreichischer Wissenschaftstheoretiker und Logiker zwischen 1933 und 1945," *Berichte zur Wissenschaftsgeschichte* 7 (1984): 227–256.

84. A useful analysis is given by S. S. Schweber, "The Empiricist Temper Regnant: Theoretical Physics in the United States, 1920–1950," *Historical Studies in the Physical and Biological Sciences,* 17 (1986): 55–98, with attention to Bridgman and Kemble; Albert Moyer, "P. W. Bridgman's Operational Perspective on Physics," *Studies in History and Philosophy of Science,* 22 (1991): 237–258, 373–397.

85. Published in *Science,* 89 (1939): 179.

86. Information and quotations in this section are taken from the following folders in the Harvard University Archives: Philipp Frank, Percy Bridgman, Edwin C. Kemble, and Harlow Shapley. I thank Clark Elliott and his staff for help with locating documents, and Kristin Peterson and Keith Anderton for much detective work in the archives.

87. Originally a lecture delivered 1 Mar. 1931 under the auspices of the Ernst-Mach-Verein; expanded into an article, "Die physikalische Sprache als Universalsprache der Wissenschaft," in *Erkenntnis,* 2 (1932): 432–465; and issued in English translation with an introduction by Max Black as *The Unity of Science* (London: Kegan Paul, Trench, Trübner, 1934). Carnap confided in a letter to von Mises, dated 19 Jan. 1934, that he was trying to find a way to spend a year at Harvard and Princeton to speak with scientists and philosophers there about common problems: Richard von Mises folder, Harvard University Archives.

88. Bessie Zaban Jones, "To the Rescue of the Learned: The Asylum Fellowship Plan at Harvard, 1938–1940," *Harvard Library Bulletin,* 32 (1984): 204–238. Shapley was intensely active also on behalf of von Mises, Tarski, Zilsel, Freundlich, and many others.

89. Frank Lentricchia, "Philosophers of Modernism at Harvard, circa 1900," *South Atlantic Quarterly,* 89 (1990): 787–832; and Feigl, "Wiener Kreis in America" (cit. n. 22), pp. 660–661.

90. Philipp Frank, *Proceedings of the American Academy of Arts and Sciences,* 80 (1951): 7–9 ("Published in cooperation with the Institute for the Unity of Science").

91. Lewis A. Coser, *Refugee Scholars in America: Their Impact and Their Experiences* (New Haven, Conn.: Yale University Press, 1984), p. 306.

92. Joergensen, "Development of Logical Empiricism" (cit. n. 62), p. 932.

2

More on Mach and Einstein

The previous chapter has documented the profound impression Ernst Mach's writing and point of view had on many scientists and intellectuals. In a declaration typical of many others, Otto Neurath wrote to Mach:

> It was this conception [the Mach Principle] in your book *Mechanik,* which never left me ever since my first reading of it, and which, on my own development of ideas . . . had its influence in a curiously circuitous way. It is the tendency to derive the meaning, the sense of individualities, from the whole, not the whole as a summation of individualities . . . With respect to your work, I have always had a deep feeling of gratitude. Through your thought processes in physics, I learned not only the advancement in the field of physics itself, but even more progress in other directions.[1]

As we saw in Chapter 1, Mach in turn was not only glad to keep in contact with his admirers and correspondents but encouraged and even pursued sympathetic readers, often enlisting them in his fight with more established opponents. In the European tradition, he was a builder of a system of thought as well as of a network of followers. Indeed, it is the combination of both of these that help make him such a significant figure in the history of modern thought.

Mach's influence on Albert Einstein during Einstein's period of greatest inventiveness and his expectations of Einstein were part of that story. There is little controversy on that point, and the existing documentation is ample, not only through an analysis of Einstein's early work but also, for example, in Einstein's surviving correspondence to Mach, starting in 1909; remarks in Einstein's letters to Michele Besso (6 January 1948), Carl Seelig (8 April 1952), Moritz Schlick (14 December 1915), and Mileva Marić (10 September 1899); as well as explicit passages in Einstein's *Autobiographical Notes* and other writings.

Moreover, there was unbroken agreement between Mach and Einstein on certain fundamental thematic notions—including the importance given to the search for unity among scientific phenomena as well as among the branches of the sciences themselves, and in their evolutionary rather than revolutionary model of scientific advance.[2] Thus, even while Einstein is the first seriously to call for deep-going modifications of the foundations of both mechanics and electrodynamics, he says that "it will only be a matter of a *modification* of our present theories, and not a complete *abandonment* of them."[3]

In all publications issued before his death in 1916, Mach seemed to reciprocate Einstein's expression of agreement on scientific matters, and he appeared to be well disposed especially toward Einstein's relativity theory. Indeed, Mach's friends and admirers tended to ascribe to Mach some part in the paternity of relativity, as indicated for example in Ferdinand Braun's letter nominating Mach for the Nobel prize, noted at the beginning of Chapter 1. To be sure, one realizes in retrospect that Mach's own comments on the record were few, brief, and somewhat cryptic. Still, one cannot blame Einstein for the shock of disappointment he (and others) experienced when five years after Mach's death there appeared over Mach's name, as the preface of Mach's posthumously issued book *Die Prinzipien der physikalischen Optik*,[4] the publication of Mach's vehement rejection of relativity. The most essential portion of that preface runs as follows:

> I am compelled, in what may be my last opportunity, to cancel my views [*Anschauungen*] of the relativity theory. I gather from the publications which have reached me, and especially from my correspondence, that I am gradually becoming regarded as the forerunner of relativity. I am able even now to picture approximately what new expositions and interpretations many of the ideas expressed in my book on Mechanics will receive in the fu-

> ture from this point of view. It was to be expected that philosophers and physicists should carry on a crusade against me, for as I have repeatedly observed, I was merely an unprejudiced rambler endowed with original ideas, in varied fields of knowledge. I must, however, as assuredly disclaim to be a forerunner of the relativists as I personally reject the atomistic doctrine of the present-day school, or church. The reason why, and the extent to which I reject [*ablehne*] the present-day relativity theory, which I find to be growing more and more dogmatical, together with the particular reasons which have led me to such a view—considerations based on the physiology of the senses, epistemological doubts, and above all the insight resulting from my experiments—must remain to be treated in the sequel [a sequel which was never published].[5]

The historian of science is confronted here by an interesting puzzle. What could have happened between the time when Einstein declared himself a "pupil" of Mach (in his letter of 17 August 1909)—and was rewarded for his declaration of allegiance by Mach's gift of a book and complimentary-sounding comments—and the composition in 1913 of this violent cancellation of Mach's earlier views? The question urges itself on us not only because the differences between these two men may illuminate the rival options for standards of good scientific practice, but also because the story of their disagreement can illustrate the state of understanding of relativity among scientists in the early years of the modernization of physics.

I have tried earlier to find plausible causal links that might help explain Mach's conversion.[6] My purpose now is to reexamine and expand the account on the basis of materials that have become available in the meantime, including two books with very different, indeed diametrically opposite, stances on the matter. These are Gereon Wolters, *Mach I, Mach II, Einstein und die Relativitätstheorie: Eine Fälschung und ihre Folgen,*[7] and a collection edited by John Blackmore and Klaus Hentschel, *Ernst Mach als Aussenseiter: Machs Briefwechsel über Philosophie und Relativitätstheorie mit Persönlichkeiten seiner Zeit.*[8] The first is in the fashion of aggressive revisionism, where some of the most crucial documentation is absent; the other is chiefly in the category of the old-fashioned presentation of actual documents, with a minimum of editorializing. Wolters's 474-page volume, originally an inaugural dissertation, centers on the startling claim that Mach's forcefully written rejection of relativity theory was a forgery committed by Mach's son Ludwig. This novel interpretation requires detailed attacks on much that has been previously written by almost all scholars dealing with the relation between Mach and Einstein (though in passing there are some plausible and useful arguments for revising certain details, such as the likely date of Einstein's single visit to Mach, or that of one of the letters between them). As we shall see, not until the last pages of the book does it become evident what may have chiefly motivated the author of this deliberately controversial work.

Wolters begins on familiar ground. Building on his interesting article "Topik der Forschung,"[9] in which he had adopted the concept of themata for the analysis of scientific thought and its development, he devotes the first chapter to the demonstration that "the influence of Mach on Einstein's development of the relativity theory may be described as a thematic influence on the first order" (p. 14). But the rest of the book, and its main novelty, is the argument that Mach, to the end of his life, far from rejecting relativity theory, was "friendly and hopeful" toward it and had "not the slightest reason" to act otherwise (ibid.).

At first glance, a lengthy concern with Mach's late disavowal of relativity seems a curious preoccupation. In a sense, it really makes little difference to history or philosophy of science whether or not the ever-skeptical but ever-surprising giant did or did not turn his back on relativity in his last years, despite the few earlier bows toward it. Except for causing disappointment, the disavowal made little difference in the long run to Einstein himself, for when it was published in 1921, Mach had been dead for five years, and Einstein had long passed beyond the stage in his evolving epistemology where Mach's interest or approval was of practical interest to him or to anyone else.

Still, the inherent sensationalism of the claimed circumstances of the supposed forgery—Wolters himself, at the Sesquicentennial Conference on Mach at Prague in 1988, called it a "Schmutzwerk" and a "soap opera" and charged the designated culprit Ludwig (dubbed Mach II) with having led a secret life of duplicity, diversion of research funds, false claims to a doctorate, psychological lability, drug addiction, and more[10]—can have a redeeming aspect. It might draw attention to a set of larger problems, that of the historic circumstances favoring the reception or rejection of new theories.

This is just where a source such as the Blackmore-Hentschel volume (which contains many of the letters quoted in this essay) becomes particularly valuable. For the letters between Mach and his correspondents allow us to trace how he first became worried about the implications of the relativity theory.[11] In this correspondence one sees Mach's growing concern, from 1909 on, with understanding the rapidly evolving conceptions. Hermann Minkowski's widely discussed lecture, *Space and Time,* of 1908 had appeared in print shortly after Minkowski's death (January 1909). As noted, Mach had been attracted to some features of relativity earlier; but after the appearance of Minkowski's paper, which reinterpreted relativity in terms of four-dimensional geometry, he clearly felt the need to understand better the mathematically complex developments of the growing theory. He was then at age 71, for many years paralyzed on his right side and suffering from many other illnesses; but he tried to keep up with his large correspondence and other involvements—not least the painful fight with Max Planck that had been launched in a lecture by Planck in December 1908 and quickly grew into vigorous published attacks on Mach's ideas, questioning even Mach's ability to serve the unity of the world picture.

On 28 March 1909, Mach wrote to Friedrich Adler, a friend also of Einstein, to indicate that he had been trying to obtain a copy of Minkowski's talk. Others of the Mach circle were also being mobilized with some urgency; Mach wrote Adler again on 16 November 1909 that his long-time disciple Joseph Petzoldt "is trying to find a man who can present the Einstein-Minkowski conceptions in a simple manner even for non-mathematicians. So far, he has not been able to find anyone. Perhaps you can direct him to somebody." Mach himself evidently could not.

Here we must stop to note a very important point that has been little discussed. It is that Ernst Mach knew and confessed he had only a rather elementary knowledge of mathematics. More than at any other time this problem was on Mach's mind in those years. He refers to it in letters to Hugo Dingler of 25 July 1911 and again on 20 November 1912, where he writes about the "weakness of my mathematical education as a youth, which unfortunately I have found no opportunity to repair."

Not until September 1910 did Mach receive, via Petzoldt, an article on relativity by J. Classen,[12] which was supposed to be of some use to Mach. Reading it, however, one sees that it did not even mention Minkowski's work. Worse, it dwelled heavily on Planck's widely noted, favorable reception of Einstein's ideas (to be discussed in Chapter 3) and even merged the two in such phrases as "the revolutionary new material in the Einstein-Planck presentation." Moreover, Petzoldt, the trusted disciple, had sent the article to Mach without any endorsement, and he added that while he thought Einstein's "fundamental idea is quite excellent," he questioned "whether he has freed himself altogether from the absolute." For example, Petzoldt said, he did not see why c and c' should be equal. (Petzoldt and others, claiming to act in Mach's spirit, objected to the idea that through the postulate of a universally constant velocity of light an "absolute" was being smuggled back into physics.)

Petzoldt remained a skeptic in a letter to Mach of 1 June 1911, in which he made a very revealing remark: "You wrote me in your last letter [one that has not been found] that the relativity principle appears to you to lack much from the point of view of epistemological criticism. I believe this also." At the least, we must infer that Mach had realized he was having real reservations about relativity.

In the meantime, Adler had also been trying to be helpful. He had informed Mach on 28 November 1909 that Einstein himself would soon publish[13] "an extensive explication in which the mathematics was insignificant [*unwesentlich*]." Mach had thanked him at once cordially for this information (3 December 1909) and on 21 February 1910 had asked again for the exact reference to it. A little earlier, on 11 January 1910, August Föppl, also an acolyte, had written to Mach, apparently also in reply to an inquiry about the Einstein-Minkowski thesis; Föppl said cautiously he had not yet formed a judgment about it but regarded it as an unverified hypothesis, and he even doubted it was in principle verifiable—precisely a criterion that empiricists and positivists regarded as the earmark of abhorrent, "metaphysical" conceptions to be avoided in good science. Mach's circle was evidently skeptical and anxious about relativity, and his own search for authoritative illumination was not going well.

It was in this setting, as we noted in Chapter 1, that the first known personal meeting occurred between Mach and a knowledgeable informant—young Philipp Frank. In his biography of Mach, Blackmore showed[14] that Frank, a Privatdozent at the University of Vienna, had first been recommended to Mach in a response of 5 June 1910 from the physicist Gustav Jäger, to whom Mach had turned in puzzlement on another problematical theory in physics (one by Paul Gerber). Frank was praised as "a man who, I think, is the best qualified of the Vienna physicists to render a judgment on the matter." Eight days later, Philipp Frank gave his opinion to Mach in a letter (dismissing Gerber's work) and then visited Mach. As Frank later reported, Mach "especially wanted to have more specific information about the application of four-dimensional geometry . . . He requested that I supply him with a printed or handwritten statement [of my ideas]. I did that . . ."[15]

At last, Mach had found his man. Another letter from Frank to Mach[16] shows us details of the growing relationship. Frank wrote obligingly: "I would like to mention further that I am now working on a representation of the theory of relativity which is understandable to non-mathematicians, as you requested in your letter, Herr Hofrat, and as Herr Professor Lampa [another of Mach's associates, at Prague University] has also asked for. I will especially try to represent Minkowski's thoughts on space and time in an understandable way."

Whether Mach's scruples were resolved in favor of Einstein-Minkowski at that time, and if so, how fully or for how long, we do not know directly from Mach. Frank himself informed Friedrich Herneck[17] he "had the impression at the time" that Mach accepted Einstein's special theory of relativity and its philosophical basis, and even that it had been Frank's own interpretation "with which

Mach agreed." Indeed, if we look at Frank's published papers on relativity around that time,[18] we watch a masterful presenter of the elements of special relativity theory—but also a conciliator; for Frank stressed the continuities with pre-Minkowskian sensibilities, for example, avoiding the use of the telltale square root of negative quantities involving *t*, which had startled so many who had encountered Minkowski's work directly.[19]

Moreover, Frank underlined that he had fashioned this article for readers "who do not master modern mathematical methods," to show that Minkowski's work brings out the "empirical facts far more clearly by the use of four-dimensional world lines." Thus Frank managed to make it appear that Minkowski's treatment preserved a science that based itself not only on a functional and operational interconnection of space and time but—fully in accord with Mach's own views—also on the primacy of ordinary, "experienced" space and time in the description of phenomena. Thus he tried to dissipate Minkowski's threat that, as Minkowski himself had put it in the first paragraph of his essay, "Space by itself, and time by itself, are doomed to fade away into mere shadows, and only a kind of union of the two will preserve an independent reality." Frank's version also kept quiet about the barely disguised idealistic boast at the end of Minkowski's article that his point of view would "conciliate even those to whom the abandonment of long-established views is unsympathetic or painful, by the idea of a pre-established harmony between pure mathematics and physics."[20]

Perhaps as a result of Philipp Frank's presentation, Mach mentioned the names of Lorentz, Einstein, and Minkowski in replying in 1910 to Planck's attack. But as is so typical in all Mach's published remarks on relativity and on those who fashioned it, the comment is cautiously short, vague, and by no means an endorsement. He simply names, in passing, these three as "physicists who by-and-by are moving closer" to the problems of matter, space, and time.[21] Similarly, in his 1909 reprinting of his 1872 work, *History and Root of the Law of Conservation of Energy,* Mach had added a reference to Minkowski's lecture of 1908, but without any comment on it and in a context that makes little sense physically but seems to be in the service of a priority claim; Mach's addition refers to general, epistemological passages of his 1872 work in which, he now explained, "Space and time are not conceived as separate entities, but as forms of mutual connection among the phenomena. Thus I am heading toward the Principle of Relativity, which is also adhered to in the *Mechanics* and the *Theory of Heat* . . ."[22] The plain reference to Minkowski can also be read as a caution; for in another of his few addenda of 1909, Mach writes: "Spaces of many dimensions seem to me not so essential for physics. I would only uphold them if things of thought such as atoms are maintained to be indispensable, and if, then, also the freedom of working hypotheses is upheld."[23]

Wolters, however, never has a moment's hesitation in reading into every phrase an endorsement. For example, as part of his systematic attempt to discredit any thought that Mach may, after all, not have been enchanted by relativity in the last years of his life, Wolters argues at length that Mach could easily have accepted Einstein's theory in the form presented by Minkowski, because even in Minkowski's formulation the basic operational meaning of time and space intervals, taken separately, are those of conventional mechanics. However, this interpretation overlooks elementary facts of physics and of history.

First, even though Frank had taken care not to stress it, it had to dawn eventually on any reader of Minkowski's essay that the basic invariant interval, the "timelike vector element" *ds,* which Minkowski defined by

$$\sqrt{c^2dt^2 - dx^2 - dy^2 - dz^2},$$

is by no means on the same operational level as the space intervals and the time intervals defined in the ordinary world of meter sticks and clocks. The negative sign under the square root signals of course that *ds* contains imaginary components. It was a central insight of Minkowski to introduce in his 1908 paper a new quantity in place of ordinary time *t,* namely, the expression $\sqrt{-1} \times t$. The alarming implication this had at the time was made clear even in Minkowski's very next sentence: "Thus the essence of this postulate may be clothed mathematically in a very pregnant manner in the mystic formula 3×10^5 kilometers = $\sqrt{-1}$ seconds."

If the ever-skeptical Mach did not sooner or later perceive in Minkowski's article that Minkowski's space-time conception was indeed an attack on the very root of a sensations-based physics, Einstein personally saw to it by sending him on or shortly before 25 June 1913 what he called in his letter to Mach of that date "my new work concerning relativity and gravitation, which has now at last been finished after infinite labors and painful doubts." It was the first extensive version of general relativity, which Einstein had begun in 1908, pursued during his years in Prague (1911–12), and now advanced in collaboration with the mathematician Marcel Grossmann.[24] It is a complex, lengthy, mathematically advanced publication based on the tensor calculus. Reading the article will quickly force one to the conclusion that Mach, who had regarded himself as an innovator along "relativistic" lines in mechanics but deplored his weakness in mathematics, would have come away at best with a feeling of increased helplessness concerning the fast-growing theory.[25] If he understood anything in it, it would have been that the theory's contact with "sense experience," so clear in Einstein's first paper of 1905, had become most tenuous.[26]

This is of course a crucial piece of the puzzle, if only because the angry rejection of relativity appearing over Ernst Mach's name was dated "July 1913," only a few weeks or days after Einstein's mailing. In the Einstein-Grossmann article, the principle of relativity was extended to apply to coordinate systems in nonuniform motion, and hence the inclusion of nonlinear transformations became necessary. As Einstein himself confessed, this step "was inevitably fatal to the physical interpretation of the coordinates . . . It could no longer be required that coordinate differences should signify direct results of measurement with ideal scales or clocks. I was much bothered by this piece of knowledge."[27] To put it in modern terms, Einstein was forced to see that "a physical significance attaches not to the differentials of the coordinates, but only to the Riemannian metric corresponding to them."[28] It was for him a wrenching experience during this part of his pilgrimage toward what he called "rational realism." As Einstein later told his old friend Cornelius Lanczos (letter of 24 January 1938): "Coming from skeptical empiricism of somewhat the kind of Mach's, I was made, by the problem of gravitation, into a believing rationalist, that is, one who seeks the only trustworthy source of truth in mathematical simplicity."

Against this background the content of the "July 1913" preface to Mach's *Principles of Optics* becomes quite plausible. To Wolters, however, ascribing this rejection of relativity to Ernst Mach is a canard which he charges was invented in an anti-positivistic "campaign," starting in the 1960s, to portray Mach unjustly as a *"philosophische Dummkopf"* (p. 403). Mach's preface would have had to be a "pathological break in Mach's personality" (p. 405) or, more probably, a forgery. Wolters's finger points to Ludwig even while he admits there is no confession and no clear motive for this alleged crime, though one may speculate about the influence of a friend, the philosopher Hugo Dingler, who, Wolters says, had against relativity a "blind hatred" (p. 402). Also, no documents seem to be available for independent scholarly study of Wolters's conjectures.

Although Ludwig repeated later (as in his letter to Petzoldt, 29 June 1920) that the abjuration was his father's, technically Ludwig had the opportunity, after his father's death in 1916, to write or rewrite some or all of the preface that bears his father's name; to incorporate in it unauthorized inventions that were contrary to his father's views; and to take the secret of his misdeeds to his grave. On the other hand, Wolters confesses at the end of his book that, after all, "it could be that the *Optik*-preface was written by Ernst Mach, or at least . . . in accord with his opinion [*in seinem Sinne*] . . . Truly, in this world much is possible" (p. 405).[29] At any rate, convinced that Ernst Mach's reputation in philosophy somehow rests to a great extent on whether he did or did not write these passages, Wolters pleads that the "antipositivism" which he says has been maligning Ernst Mach should from now on target the son instead. Thus ends the book.

Wolters would object at this point that two significant documents have not been mentioned: Einstein's undated letter to Mach (probably written at the end of 1913 or the beginning of 1914) and Mach's letter to Petzoldt dated April 1914. A few words in these are the centerpieces of Wolters's "evidences" that Mach could not, in July 1913, have been hostile to Einstein's work. (Passages in one other letter, Mach to Petzoldt, 1 May 1914, could be stretched to hint in that direction, but at least three other documents say as much in opposition.) Let us therefore finally look at these two in turn.

The first sentence in Einstein's undated reply (of late 1913 or early 1914) says merely, "I am very glad about your friendly interest which you bring toward the new theory." Mach's own letter, to which this sentence refers, has not been found. But let us assume that Mach wrote some explicitly pleasant phrases to this ascending star, who had once called himself his "admiring student," and even assume further that such phrases referred to the Einstein-Grossmann paper. It would not have been the first or the last time that a scholar made polite remarks about a mailing that he really neither understood in detail nor approved of in general. In fact, one finds examples of this sort right in Mach's letters to others, particularly to younger scientists of obvious quality who presented themselves as admirers. For example, as we already noted in Chapter 1, Mach was generally supportive of William James in direct correspondence with him but did not mind criticizing his work in letters to others.[30] Therefore it is dangerous to let so much of a complex case rest on what can be read into Einstein's two words, "friendly interest."

This argument becomes even more compelling as soon as one continues the quotation; for in his letter to Mach Einstein added: "The mathematical difficulties against which one comes up in following these thoughts were unfortunately also for me very great." Mach appears to have made a comment on the grave mathematical complexities of the work, which in fact had necessitated Einstein's seeking the help of the mathematician Grossmann.

Wolters's other key piece of evidence is in Mach's reply to Petzoldt of April 1914. At least there we are dealing with Mach's own words. But there, too, one cannot read into them an acceptance of Einstein's relativity theory. Mach writes that he is pleased with Petzoldt's article,[31] sent to him by its author, first because "you have richly honored my modest merits with regard to this theme." (In fact, Petzoldt's article flatters Mach on page after page, in the sycophantic way Mach had come to expect from his followers.) Then Mach adds the vague and weak phrase that Wolters takes to be a decisive indicator of acceptance: Mach says he likes the article *"auch sonst"* [also otherwise, or in general].[32] So the obvious thing to do, one would have thought, is to find out what lies behind the phrase *auch sonst*—that is, to read Petzoldt's article, in order to see what kind of relativity theory Petzoldt is writing about in 1914. Indeed, anyone who toys with the idea of accepting Mach's opaque two words as an endorsement of relativity is condemned actually to study that article: a 56-page piece, in the official organ of the recently founded Society for Positivistic Philosophy, written by its founding chief and member of its editorial board.

This article further opens the window both on the standards of acceptable science and on the level of understanding of physics then current in Mach's circle. In fact, Petzoldt's piece turns out to be not a report on relativity in the form it had then reached, but a quasi-philosophical attempt to show that Petzoldt's own elementary and idiosyncratic version of relativity theory fits with the "relativistic positivism founded by E. Mach and R. Avenarius." The main problem discussed by Petzoldt is the relation of the physical and the psychological basis of experience, chiefly from Mach's point of view. The rejection of absolute space by both Mach and Einstein is highlighted, and the work of the latter is praised for showing that the Lorentz-Fitzgerald contraction is not a phenomenon requiring a physical "explanation" but follows merely from a proper description.

The important point Petzoldt wants to make here is "that physics in the end can deliver no more than the *description* of events; that complete description *is complete explanation;* and that at bottom physics has never delivered anything else" (p. 16; emphasis in original). Here Minkowski's essay is briefly said to be pointing to the same conclusions. Minkowski is occasionally referred to also later, but mention of his work is followed by the observation that this "marvelous" theory depends, as Einstein's does, entirely on the presupposition of the universal constancy of light velocity and, like previous theories, may well fail with the eventual failure of that presupposition. What will then remain standing is Mach's anti-absolutistic theory, "for it rests on sense-physiological foundations."

Moreover, Petzoldt warns that the relativity of Einstein-Minkowski must be put on a more visually intuitive, *"anschauliche"* ground. And while Einstein-Minkowski's four-dimensional world is acceptable as long as it is considered as "merely a conceptual system," Petzoldt cannot agree with Minkowski's conclusion that the four-dimensional world is given through the events that take place in it.

Petzoldt points out other "deficiencies" also. A few of these will suffice to indicate the general tone. Einstein's own formulation of the relativity principle is charged to be on the level of a "naive assumption, not of an assumption after critical examination." Einstein's conception of "natural law," too, has an element of arbitrariness; the difference between space-time determinations and natural laws cannot be tolerated. Therefore, "we see how important epistemological studies can be for a theoretical physicist and mathematician." Properly understood, relativity can teach what is meant by a "real" versus an "apparent" change, as in the length of a rapidly moving body or even in the observed shapes of objects as our vantage point is changed. Thus, observed shapes are not appearances based on hidden "things-in-themselves" but "only psychobiological functions." At its deepest level, understanding itself consists "in the relativity of all actuality and its substancelessness" (p. 46).

Out of the blue, there appears at this point—again only in the most elementary way—a discussion of the recently popularized clock paradox of relativity theory. Now Petzoldt reveals his ignorance in plainest terms; for he claims that while the traveler, on returning to his home-based fellow creatures, will seem to the latter to have aged less, the traveler himself will have a *symmetrical* experience and will perceive the home-based fellows to have remained more youthful than he is himself! To think otherwise, says Petzoldt with solemn authority, is "an error and a regression back into absolutistic thinking" (p. 50).[33]

One additional point is noteworthy in Petzoldt's paper, a note inserted without clear context just before his doomed discussion of the clock paradox. It is a reference to two of Einstein's papers extending relativity theory to include the "gravitation problem," one of which is the paper by Einstein and Grossmann. Petzoldt's text does not make clear why they are cited; for instead of attempting to discuss or describe the papers, he says merely, in a way perhaps meant to express his disapproval: "Ordinarily the physicist in his *Anschauungen* does not hew closely to his formulas, but weaves around them a multiplicity of considerations which have no hold in the formulas themselves." And Petzoldt adds that "it is the function of epistemological critique, as carried out by Mach, to discover these addenda and their insupportability."

One can only conclude that if Mach—quite apart from being flattered—had indeed found pleasure and agreement with Petzoldt's 1914 paper, as his letter to Petzoldt indicates, it would necessarily have to be pleasure in and agreement with Petzoldt's multifarious *objections* to Einstein's relativity. More important, Mach's agreement with Petzoldt's version of relativity would be a terrible indictment of Mach's ignorance about the current state of relativity theory. In sum, on this cumulative evidence—alas—it no longer matters *who* wrote Mach's disavowal dated July 1913. Whether he intended to accept it or reject it, Ernst Mach at that point no longer knew what relativity was about.

Notes

1. O. Neurath to E. Mach, no date [probably 1915], in Joachim Thiele, *Wissenschaftliche Kommunikation: Die Korrespondenz Ernst Machs* (Kastellaun: A. Henn Verlag, 1978), p. 100.

2. For example, A. Einstein, "Zum gegenwärtigen Stand des Strahlungsproblems," *Physikalische Zeitschrift,* 10 (1909): 185–193.

3. Einstein, "Zum gegenwärtigen Stand," p. 192. Emphases in original.

4. Leipzig, 1921. Available in an English translation of 1926 as E. Mach, *The Principles of Physical Optics* (New York: Dover Publications, Inc., no date), pp. vii–viii.

Einstein and others soon forgave Mach for his sharp rejection of the relativity theory, considering it a result of age and/or illness.

5. From the 1926 translation in the Dover edition; I have corrected some mistranslations.

6. In G. Holton, *Thematic Origins of Scientific Thought: From Kepler to Einstein* (Cambridge, Mass.: Harvard University Press, 1973), chapter 8, and in the revised version of 1988, chapter 7.

7. Berlin, New York: Walter de Gruyter, 1987.

8. Vienna: Wilhelm Braunmüller, Universitäts-Verlagsbuchhandlung GmbH, 1985.

9. Gereon Wolters, "Topik der Forschung," in C. Burrichter, R. Inhetveen, and R. Kötter, eds., *Technische Rationalität und rationale Heuristik* (Paderborn: Schoningh, 1986), pp. 123–154.

10. Elaboration on these points was provided in Gereon Wolters, "Atome und Relativität—Was meinte Mach," in R. Haller and F. Stadler, eds., *Ernst Mach—Werk und Wirkung* (Vienna: Verlag Hölder-Pichler-Temsky, 1988).

11. The 128 letters, obtained by Blackmore and Hentschel since 1981, are thus complementary to Joachim Thiele's 1978 volume cited in n. 1. An additional set of documents has been recently published as Part 2 (pp. 167–305) of the book by Haller and Stadler cited in n. 10.

12. J. Classen, "Über das Relativitätsprinzip in der modernen Physik," *Zeitschrift für Physikalischen und Chemischen Unterricht,* 23 (1910): 257–267. I am grateful to Dr. Dieter Hoffmann for his help in finding a copy of this publication.

13. A. Einstein, "Le principe de relativité et ses conséquences dans la physique moderne," *Archives des Sciences Physiques et Naturelles,* 29, no. 1 (1910): 5–28, 125–144. The level is indeed very elementary; and Minkowski's work is barely touched on, in three paragraphs.

14. John T. Blackmore, *Ernst Mach, His Work, Life, and Influence* (Berkeley, Calif.: University of California Press, 1972), p. 263.

15. Friedrich Herneck, "Die Beziehungen zwischen Einstein und Mach dokumentarisch dargestellt," *Wissenschaftliche Zeitschrift der Friedrich-Schiller-Universität Jena, mathematisch-naturwissenschaftliche Reihe,* 15 (1966): 7. See also Blackmore, *Ernst Mach,* p. 183, for the comment that "Ernst Mach sent Frank two letters in 1910 at least partly in the hope that Frank could help clarify the ideas of Einstein and Minkowski for him."

16. Given by Blackmore, *Ernst Mach,* p. 264, as undated "but clearly also in 1910."

17. Herneck, "Beziehungen zwischen Einstein und Mach," p. 7.

18. For example, in *Zeitschrift für Physikalische Chemie,* 74 (1910): 466–495, which Herneck (ibid., pp. 7, 13) says is the most relevant, and of which Frank sent a reprint to Mach.

19. Doing so was a point of pride for Frank, as he once told me. Moreover, in this respect he was following Einstein's lead in his 1910 survey article, cited in n. 13.

20. For a discussion of the essentially Platonic undercurrent in

Minkowski's article, see Peter L. Galison, "Minkowski's Space-Time: From Visual Thinking to the Absolute World," *Historical Studies in Physical Science,* 10 (1979): 85–121.

21. E. Mach, "Die Leitgedanken meiner naturwissenschaftlichen Erkenntnislehre und ihre Aufnahme durch die Zeitgenossen," *Physikalische Zeitschrift,* 11 (1910): 605.

22. Translated from Joachim Thiele, ed., *Ernst Mach* (Amsterdam: E. J. Bonset, 1969), p. 60. The book contains a reprinting of the original German edition of Mach's *Die Geschichte und die Wurzel des Satzes von der Erhaltung der Arbeit* (1872) and Mach's brief addenda to the second printing (1909).

23. Quoted in Thiele, *Mach,* p. 59.

24. A. Einstein and M. Grossmann, "Entwurf einer verallgemeinerten Relativitätstheorie und einer Theorie der Gravitation," *Zeitschrift für Mathematik und Physik,* 62 (1914): 225–261. The title page of the journal says "Issued on 30 January 1914"; but a separate, 38-page publication of the article had appeared earlier (Leipzig: B. G. Teubner, 1913). Wolters agrees that this article was the one Einstein mentioned in his letter of 25 June 1913 as having been sent to Mach.

25. Einstein had given a preview of that part of the article as a talk in Vienna during the 85th Naturforscherversammlung in Vienna. See *Physikalische Zeitschrift,* 14 (1913): 1249–1266.

This lecture may also have been the occasion of Einstein's visit to Mach, during which apparently the main subject of discussion was their difference concerning the assumption of the existence of atoms. The visit is described in Philipp Frank, *Einstein, His Life and Times* (New York: Alfred A. Knopf, 1947), pp. 103–105.

26. Moreover, more evidence was now accumulating that Einstein was being captured by—of all people!—Max Planck, no matter how Einstein protested privately to Mach, in two of his letters, about Planck's attacks on Mach. It could not have helped matters that the Einstein-Grossmann article refers to a Planck publication as if it were the exemplary exposition of "the customary relativity theory" (p. 226); that Einstein published a glowing article about Planck in 1913 in the first volume of the new *Naturwissenschaften;* and that he was negotiating with Planck to leave Zürich and join him in Berlin.

27. A. Einstein, "Notes on the Origin of the General Theory of Relativity," lecture of 20 June 1933 at the University of Glasgow, reprinted frequently, e.g., in Albert Einstein, *Ideas and Opinions* (New York: Dell Publishing Co., 1954), p. 281; see also Einstein's *Autobiographical Notes,* p. 67.

28. Einstein, *Ideas and Opinions*, p. 282.

29. He might also have added, as others have shown, that the wording in the preface, and other internal evidence, shows it to be in accord with Ernst Mach's idiosyncratic expression, style, and ideas. See John Blackmore, "Mach Competes with Planck for Einstein's Favor," *Historia Scientarium*, 35 (1988): 45–89, and John Blackmore, "Mach über Atome und Relativät—Neueste Forschungsergebnisse," in R. Haller and F. Stadler, *Ernst Mach* (cit. n. 10).

30. Mach to Anton Thomsen, 4 September 1909. Mach writes he likes James's *Principles of Psychology* well, but also criticizes James's proclivity to romanticism and spiritualism. Similarly in Mach's letter to Thomsen of 21 January 1911, on James's "dangerous argument."

31. Joseph Petzoldt, "Die Relativitätstheorie der Physik," *Zeitschrift für positivistische Philosophie*, 2 (1914): 1–56.

32. Petzoldt is another case where Mach rarely expressed an objection directly to his face but did so in letters to others, e.g., to Friedrich Adler, on 20 August 1909 and 23 January 1910.

33. Klaus Hentschel, *Interpretationen und Fehlinterpretationen der speziellen und der allgemeinen Relativätstheorie durch Zeitgenossen Albert Einsteins* (Boston, Berlin: Birkhäuser Verlag, 1990), pp. 101–420, discusses Petzoldt's subsequent, mostly unsuccessful attempts to understand the special and general relativity theories; it includes the correspondence with Einstein, who eventually concludes with the remark, "your misunderstanding is quite fundamental" (p. 416).

3

Quanta, Relativity, and Rhetoric

By Way of Prologue

Rhetoric in Science? To a scientist, the very phrase has all the signs of an oxymoron. Since ancient times rhetoric has been essentially the art of persuasion, in contrast to the art of demonstration. Of all the claims of modern science, perhaps the strongest is to have achieved, in painful struggle over the past four centuries, an "objective" method of demonstrating the way nature works, of finding and reporting facts that can be believed regardless of the individual, personal characteristics of those who propose them, or of the audience to which they are addressed. This distinction of the roles of objectivity and subjectivity is clear in Aristotle's *Rhetorica:*[1] Of the three kinds of "modes of persuasion" available to the speaker relying on rhetoric, only the third "depends on the proof, or apparent proof, provided by the words of the speech itself," whereas "the first kind depends on the personal character of the speaker, and the second on putting the audience in a certain [right] frame of mind." Indeed, the chief rhetorical weapon is the speaker's inherent moral character:

> We believe good men more fully and more readily than others . . . It is not true . . . that the personal goodness revealed by the speaker contributes nothing to his power of persuasion; on the contrary, his character may almost be called the most effective means of persuasion he possesses.

Science had to find the escape from this moralizing and personalizing mode of discourse and invent means of persuasion other than the probity or the stylistic ruses of the presenter. As if to underline that this self-denying ordinance is one of the criteria of demarcation of science, Robert Hooke's draft preamble to the original statutes of the Royal Society of London specifically disavowed that the scientists intended to "meddle" with "Rhetoric." Since about the midseventeenth century, the writings of scientists have increasingly reflected their agreement with such admonitions. Thus Newton adopted for his *Principia* a structure that suggested parallels with that exemplary model of objectivity, Euclid's presentation of geometry, and he opened the first book of his *Opticks* with the implication that the work is free from conjecture, analogy, metaphor, hyperbole, or any other device that might be identified with the rhetorician's craft. Rather, Newton says, "My Design in this Book is not to explain the Properties of Light by Hypotheses, but to propose and prove them by Reason and Experiments."[2]

The well-tested machinery of logic and analysis, the direct evidence of the phenomena—who can resist these? Who would need more? Newton and the scientists who came after liked to be considered little more than conduits through which the book of nature spoke directly, across the great divide between the independent, outer world of phenomena and the subjective, inner world of the observer. But because they are in consonance with the "Tenor and Course of Nature,"[3] their reports are free from the vagaries and limitations of mere humans. In Alexander von Humboldt's phrase, they should be the results of observation, stripped of all "charms of fancy." Or at least, as Louis Pasteur advised his students—and as is current practice in any research article submitted to a science journal—"Make it look inevitable."

Here indeed there does reveal itself a connection with the final aim of the old rhetoric. For as Aristotle noted, the most desirable of the various propositions of rhetoric is the "infallible kind," the "complete proof" (τεκμήριον): "When people think that what they have said cannot be refuted, they think they are bringing forward a 'complete proof,' meaning that the matter has now been demonstrated and completed."[4] Thus alerted, we now remember that a number of recent investigations by historians of science have shown that at least *before* a work has ripened into publication, during its nascent period, traditional rhetorical elements, such as conjecture, analogy, metaphor, and even the willing suspension of disbelief, can be powerful aides to the individual scientist's imagination.[5] Therefore it is reasonable to ask whether some of the dramatic repertoire is not, after all, used—and perhaps even necessary—in the resulting publication also.

Indeed, I shall propose here and try to make persuasive by illustration a view different from and complementary to the usual way of reading a historic scientific paper. It is this: The publication is not only the author's account of the outcome of the struggle with nature's secrets—which is the publication's main purpose and chief strength, hence the scientist's preferred interpretation—but it may also be read as the record of a discourse among several "Actors," whose interplay shapes the publication. And as we shall see, in that respect it is analogous to the script of a play in which a number of characters appear, each of whom is essential to the total dramatic result.[6]

In using the word *complementary,* I stress that I am not proposing that we may or even can choose between these two ways of reading. The second will not detract in any way from the achievement intended by the first. We shall simply be looking at the presentations of scientists not chiefly from the viewpoint of their properly intended prime audience, but as it were orthogonally, as seen from the wings. However, we must not expect that the existing published scientific work will make it any simpler to discern its internal rhetoric than it has been to derive from it the original motivation or the actual steps that led to the final result. Indeed, rare is the scientist who helps the historian or philosopher of science to penetrate beyond the mask of inevitability, to witness what Einstein called "the personal struggle," to glimpse the various influences—biographic, thematic, institutional, cultural, etc.—that gave birth to a publication.

We cannot expect otherwise, for there are good sociological reasons for that neglect and impatience. The very institutions of science, the selection and training of young scientists, and the internalized image of science are all designed to minimize attention to the personal activity involved in publication. Indeed, the success of science as an intersubjective, consensual, sharable activity is connected with the habit of silence in research publications about individual personal struggles. Hence the useful fiction that science takes place in a two-dimensional plane bounded by the phenomenic axis and the analytic axis, rather than in a three-dimensional manifold that includes the thematic dimension.[7] Moreover, the apparent contradiction between the sometimes illogical-seeming nature of actual discovery and the logical nature of well-developed physical concepts is being perceived by some scientists and philosophers as a threat to the very foundations of science and to rationality itself. (The vogue to attempt, by a "rational reconstruction" of a specific case, to demonstrate how a scientific work should have been done seems to have been so motivated.)

Still, we shall learn how to read with minutest attention what a scientific author says or does not say, look also for unstudied evidence, and instead of settling only for the surface-reading that the publication invites, peer also behind the mask of inevitability. Works of literary or political intent have been subjected to an analysis of rhetorical elements for over two and a half millennia. Now we shall begin to distinguish the corresponding elements in the discourse of and about science: in the nascent phase during which the scientists weigh the persuasiveness of their ideas to themselves; in their published results; in the debates about these; in biographical and autobiographical writings of scientists; in scientific textbooks; and also in the uses made of scientific findings in controversies—a second-order phenomenon, a "rhetoric about rhetoric."

Rhetoric of Assertion vs. Rhetoric of Appropriation / Rejection

Comparing a scientific paper with the various responses to it makes it evident that, to begin with, one must distinguish between a proactive Rhetoric of Assertion and a reactive Rhetoric of Appropriation / Rejection. The first of these expresses that about which a scientist has convinced himself or herself, and hopes to persuade others of, when writing the publishable version of the work. The second characterizes the responses to it by contemporaries and later readers—responses that, we should note, are shaped in turn by the responders' own commitments to their own Rhetoric of Assertion. The success or refusal of recognition, or its delay, as well as misplaced reinterpretation even by those who thought of themselves as converts, can thus be understood in terms of a match or mismatch between key elements in each of these two types of rhetoric.

Foremost among these key elements in many cases in the history of science are thematic commitments: those of the originator and those of the critics or opponents or would-be disciples. Since thematic commitments are not always consciously held, we are therefore often forced into a quasi-archeological task: to dig below the visible landscape of a controversy in order to find the usually invisible but highly motivating matches, mismatches, and clashes between the respective sets of themata that have been adopted by the various participants—and not only of the individual themata, but also of constellations of them that define the locally held scientific world pictures. Such correspondences and conflicts can be considered as interactions among contesting claimants in what Michel Foucault has termed "rhetorical space."[8]

Good examples for our study come from those two classic papers that, more than most others, opened the path and set the style of physical science in our century. I shall deal first with illustrative comments on Niels Bohr's seminal paper, "On the Constitution of Atoms and Molecules." Published in three parts beginning in July 1913, this paper presented the working picture of the nuclear atom with its orbiting electrons, including its spectra and some indication of its chemical properties, a picture that has long been familiar to the point of banality. A physicist of today will agree with Emilio Segrè's assessment: "The sophisticated reader will admire the dexterity with which Bohr sails across a sea full of treacherous shoals and lands safely . . ."[9]

That is, of course, how it looks to those who have been brought up on Bohr's model. But the more immediate response was captured by Leon Rosenfeld. In his introduction to the reprint of Bohr's 1913 papers, he wrote:

> The daring (not to say scandalous) character of Bohr's quantum postulate cannot be stressed too strongly: that the frequency of a radiation emitted or absorbed by an atom did not coincide with any frequency of its internal motion must have appeared to most

> contemporary physicists well-nigh unthinkable. Bohr was fully conscious of this most heretical feature of his considerations: he mentions it with due emphasis in his paper, and soon after, in a letter to S. B. McLaren (1 September 1913), he writes: "In the necessity of the new assumptions I think that we agree; but do you think such horrid assumptions, as I have used, necessary? For the moment I am inclined to most radical ideas and do consider the application of the mechanics as of only formal validity."[10]

Indeed, if one carefully reads Bohr's paper (in volume 26 of *The Philosophical Magazine*), especially Part I, finished in haste in less than three months in early 1913, it becomes clear why it had initially such a mixed reception and why Bohr, in the interviews near the end of his life, expressed some regrets about having published it in that form. To compare the Rhetoric of Assertion with that of Appropriation / Rejection, to see how differently Bohr's work appeared to the young man himself and to some of the lions, we can also make use of fairly reliable accounts of "unstudied," spontaneous, spoken comments, of which in this case there happily exist a good supply.

Abraham Pais has published a collection of typical reactions under the heading, "It was the epoch of belief, it was the epoch of incredulity,"[11] though there was at first far more of the latter. A few, notably Einstein, Debye, and Jeans, were fully receptive. But that was a distinct minority view. Thus Otto Stern told Pais that not long after the publication of Bohr's papers, Stern and Max von Laue, while on an excursion on the Uetliberg outside Zurich, swore what they called a solemn Uetli Oath: "If that crazy model of Bohr turned out to be right, they would leave physics."[12] Lord Rayleigh with lofty simplicity said of the paper, "It does not suit me." J. J. Thomson's obstinate objection to Bohr's conception was palpable in most of his writings on the atom from 1913 to 1936. H. A. Lorentz's leniency was clear, but it had its limits. As *Nature* reported in its account of the first meeting in Britain at which Bohr spoke about his atom, Lorentz (who had already objected earlier that "the individual existence of quanta in the aether is impossible") intervened to ask "how the Bohr atom was mechanically accounted for," and Bohr had to acknowledge "that this part of his theory was not complete, but . . . some sort of scheme of the kind was necessary."[13]

Bohr himself noted later,

> When my first paper came out, it was actually objected to in Göttingen. There was no interest for it, and, as I told you, there was even a general consent that it was a very sad thing that the literature about the spectra should be contaminated by a paper of that kind. The paper was just a playing around with numbers and there was nothing in it . . . It was clear that that was the general consent . . . Because at first there actually was nothing. And that's what we'll come to. But now the question is, how was it presented?[14]

That was indeed the question. In a preview of his work, Bohr had warned Rutherford in 1912 that he, Bohr, would have to adopt a hypothesis "for which there will be given no attempt at a mechanical foundation *(as it seems hopeless)*."[15] But when Rutherford actually saw the manuscript, he had to write to Bohr on 20 March 1913,

> the mixture of Planck's ideas with the old mechanics [Bohr himself had characterized it as "the delicate question of the simultaneous use" in a letter of 6 March 1913] makes it very difficult to form a physical idea of what is the basis of it all . . . How does the electron decide what frequency it is going to vibrate at when it passes from one stationary state to another?

A fair question—it took until 1917 for Einstein to show a way.[16]

What most concerned many of Bohr's readers—brought up on atom models, such as Thomson's, that were considered "mechanically accounted for"—when forced to decide on appropriation or rejection, was not only Bohr's presentation, a Rhetoric of Assertion in which he rather cavalierly mixed classical and quantum physics, but also his introduction into his atom of the thema of discontinuity as well as that of probabilism rather than Newtonian causality—antithemata with respect to the classical foundations. These foundations were threatened at the time also from other directions. Returning from the 1911 Solvay Conference, the first summit meeting on quantum physics, James Jeans had baldly stated what to many was an ominous advent in the thematic base of physics: "The keynote of the old mechanics was continuity, *natura non facit saltus*. The keynote of the new mechanics is discontinuity."[17]

But Jeans was far more ready for this profound change than many others. Eddington said of him that he was the only one in England who had been converted to quantum physics by the Solvay Conference. Henri Poincaré, returning from the same meeting, spoke for the large majority when he concluded wistfully in the last year of his life:

> The old theories, which seemed until recently able to account for all known phenomena, have recently met with an unexpected check . . . A hypothesis has been suggested by M. Planck, but so strange a hypothesis that every possible means must be sought for escaping it. The search has revealed no escape so far . . . Is discontinuity destined to reign over the physical universe, and will its triumph be final?[18]

Unlike so many of his elders, the twenty-seven-year-old Niels Bohr had built up no equity in the themata of the older physics. He was young enough to have encountered the existence of quantum ideas from his student days on. Moreover, in working on his doctoral dissertation on the electron theory of metals, which he had just completed, he had come to understand more clearly than his own examiners that the classical conceptions were simply incapable of dealing sufficiently with, for example, specific heats, or the high-frequency portion of black-body radiation, or the magnetic properties of matter.

Thus, to understand the argument by which the author of a work has convinced *himself*, one must look for roots of the argument that may have already appeared in his previous work. Bohr's 1913 paper is a point on a developing trajectory of personal science (S_1) that intersects upon publication in July 1913 with public science (S_2). On the earlier part of S_1, we find not only Bohr's doctoral thesis but his abortive discussions with J. J. Thomson during Bohr's stay in Cambridge and Bohr's productive work on alpha-particle scattering at Rutherford's laboratory in Manchester; their traces can be found on the first pages of the July 1913 paper.

Not One Actor but (at Least) Two

We generalize this point in the following proposition:

I. *A scientist's current work is likely to be the continuation of a soliloquy that has its origins in his earlier work.*

A second proposition follows as if by symmetry:

II. *In the scientist's current work one may discern evidences of the direction that his future work is likely to take.*

To add to the illustrations already given for proposition I, we may note that Bohr's courage in July 1913 is a consequence of his earlier radicalization. Rutherford's nuclear model of the atom was discovered quite unexpectedly at the end of 1910 and published in 1911. It, too, was at first widely disbelieved, and Rutherford did not insist on it himself (as indicated by his silence about it at the 1911 Solvay Conference). But its implications were enormous and were perhaps best caught in the artist Kandinsky's outburst that now that the old atom had been destroyed, the whole existing world order was annihilated and so a new beginning was possible.[19] To Rutherford's young collaborators in Manchester, especially to Bohr, who had fled there from Cambridge and its resistance to new ideas, Rutherford's discovery of the concentration of the atom's mass had revealed the crucial flaw in the then reigning model of the atom (primarily J. J. Thomson's), even though that model had, among other useful features, yielded a plausible explanation for the size of the atom and for multiple-scattering data.

In Rutherford's atom model, however, no one knew any longer what to do with the electrons around the nucleus. Thomson thought of that as "a very great calamity";[20] but when Bohr was asked, "Were you the only one who responded well to it [the Rutherford atom]?" he replied, "Yes, but you see I did not even 'respond' to it. I just believed it."[21] He had evidently been ready for it after his unsatisfactory struggle with the classically based atom during his dissertation work: "Now it was clear, and that was *the* point in the Rutherford atom, that we had something from which we could not proceed at all in any other way than by a radical change."[22] Or, as he had put it in his July 1912 "Memorandum" for Rutherford, the stability of the electrons' configuration had to "be treated from a quite different point of view."[23]

The direction in which to seek salvation was clear. For some years, Planck's quantum of action *h* had been the tool for understanding black-body radiation, and it promised to do the same for specific heats. It had a magic about it, at least for young people ready to risk it. (As Edwin C. Kemble, who initiated quantum physics research in the United States in his twenties, recalled as his own motivation: "Anything with quantum in it, with *h* in it, was exciting."[24] It was, as so often in the history of science, a matter of being ready to embrace the new themata. Even those features of Bohr's atom which to others eventually were the most persuasive—for example, the correct prediction of spectral lines and the derivation of the value for Rydberg's constant—were not essential for convincing Bohr himself; for we know now that he stumbled on these aspects only at the last minute, in early 1913, when the main parts of his paper had been fixed.

It is also easy to illustrate proposition II. Thus a striking feature of Bohr's thinking, which then suffuses all his later works, including especially that on complementarity. His radicalization had not forced him, as it might have others, to abandon entirely the old, mechanistic conception. On the contrary, he held that "by analogy" to what is known for other problems, it seemed legitimate to continue to use the old mechanics *side by side* with the new quantum physics, often within the same paragraph of the 1913 paper. This is just what Rutherford had found most puzzling. But it is at the heart of Bohr's daring proposal in 1913 of what he named later the "correspondence point of view," which in turn, from 1927 on, burgeoned into his "complementarity argument."

We can now summarize this segment: To a greater or lesser degree, a publication can be read as the extrapolation from the author's past, as well as the staging area for a future expedition. To put it differently, in studying the Rhetoric of Assertion of an author in a given work, we discern *that he disaggregates into two Actors, engaged in two different soliloquies on the same stage.*

Actor 1 is engaged in an internal dialogue with his own recent or more distant past work, out of which the new work is growing.

Actor 2 has begun to engage in thoughts that will not come to full fruition for some time in the future. The author's production results in good part from both soliloquies and receives different characteristics from each: on one side, conviction from past difficulties being now conquered; on the other side, conviction from the attractiveness of further successes that perhaps only dimly but tantalizingly beckon—especially in Bohr's case, the new thema of complementarity; the hope for a greater unification of understanding both chemical and physical properties of matter through his new atom; and the feeling that something wonderful looms beyond. Thus in his letter to G. Hevesy, Bohr writes on 7 February 1913,

> . . . I don't speak of the results which I mean that I can obtain by help of my poor means, but only of the point of view—and the hope to and belief in a future (perhaps very soon) enormous and unexpected?? development of our understanding—which I have been led to by considerations as those above.[25]

It is a near paraphrase of Galileo's prophecy, at the end of Day Three in his *Two New Sciences,* that "the principles which are set forth in this little treatise will . . . lead to many another more remarkable result." In this way, while Actor 1 is animated by the satisfaction of recent difficulties surmounted, Actor 2 is pulled forward by attraction to the greater goal on his agenda. Moreover, one of the most important long-term functions of a seminal paper in science is plainly rhetorical: that its readers come to share the author's excitement, his sense that new vistas are being opened, that new questions can be raised and perhaps answered. (The chemist Dudley Herschbach has christened it "the spiritual effect" of good new science.)

But as we also have begun to note, the two Actors are by no means alone on the stage defined by the text of the paper. Each carries on his monologue in the imagined presence of his important colleagues. The published paper bears witness to that: One can unravel which of the two Actors is speaking a line, and against the background of which other imagined supporter or opponent that line is composed. (By no means all of these will be identified in the text by name or in the notes.) Thus Bohr's paper in its first paragraph is the acknowledgment by an acolyte of "Professor Rutherford" (who also served as the identified communicator of the paper to the journal), of the motivating power of Rutherford's recently discovered nuclear atom. The next two paragraphs are a continuation of it, with the addition of a cautious acknowledgment of the power still exercised by the commanding ghost of "Sir J. J. Thomson." In the fourth paragraph, we see Bohr accepting the promise of the revolution Planck had introduced in 1900 (much against his own will). And only then, in the last half-sentence of that paragraph, in a throw-away line, the first evidence of Bohr's own ideas: a remarkable feat of confident intuition, made almost incomprehensible by the failure of the young author to articulate his own voice in that distinguished company.[26]

By page four, Bohr introduces the strange idea that the frequency of the radiation emitted in binding the electron to the atom is "equal to half the frequency of revolution of the electron in its final orbit." A typical early reaction was that this was "a crazy stunt"—but here we have again the emergence of Actor 2 on the stage, presenting an apparently unsupported argument that will develop later into Bohr's treasured correspondence argument by which he tries to hold on to both classical physics and quantum physics.

In this way, a paper can be resolved paragraph by paragraph into the main rhetorical components in the assertion stage, into, for example, the various parts that are carried by different Actors. Moreover, one can also differentiate between the rhetorical components in the subsequent stages of appropriation or rejection—which in the case of Bohr's atom was particularly turbulent for the first years among physicists in the United States, who were puzzled whether to regard Bohr's two-dimensional atom model as a discovery, an analogy to the three-dimensional nature of matter, or a powerful metaphor.[27] But instead of pursuing this further for this particular case, and to indicate the universality of the role of rhetoric despite great individual differences, I turn now to another seminal paper in the history of early twentieth-century physics.

Relativity: Its Publics and Its Authors

The case of Einstein's early writings is analogous to Bohr's chiefly in one respect: Albert Einstein's formulation of what is now called special relativity has also become so familiar to us that one may say, as he did about Ernst Mach's ideas, that one has imbibed it with one's mother's milk. Eventually, relativity theory became one of the "charismatic" activities, to use the terminology of Joseph Ben-David. Therefore it takes an act of serious will to free oneself from an ahistorical view about Einstein's claims as they were launched in a quick series of communications, starting with the publication of the first paper on 26 September 1905, "On the Electrodynamics of Moving Bodies." From the perspective of rhetoric, this paper was almost calculated to be off-putting to the typical reader of 1905. Indeed, as Einstein had predicted in one of his letters, in terms of the immediate reception by the large scientific community, this work could be regarded as a failure. Approval was certain only from the few personal friends of this unknown and sociologically "marginal" man, fellow marginals such as Michele Besso, Joseph Sauter, Marcel Grossmann, and Conrad Habicht. While Bohr's paper showed from the first sentence that he was conscious of moving, as indeed he did, in Olympian company, Einstein's emanates the sense that the young author is unused or unwilling to address himself properly to his "betters" (as indeed was also the case).

A fair understanding of Einstein's formulation grew among major physicists only slowly during the first few years. The Rhetoric of Appropriation / Rejection was heavily weighed to the latter. And even those who one by one were converted, in almost all cases, interpreted the main point of Einstein's work in a significantly different manner than he himself had intended. It is reasonable to say that it took six years, with the appearance of Max von Laue's first textbook on relativity in 1911, for an irreversible change in the unfavorable balance to be signaled; and some, including H. A. Lorentz, did not make their peace with Einstein's relativity to the end of their days. The one great exception in all this, as we shall see, was Max Planck, whom Einstein himself regarded as his first and crucially important champion among the elite. And even there, when it came to the extension into general relativity some eight years later, Einstein complained in one of his letters to Ernst Mach that Planck's "stance to my theory is also one of refusal."[28]

As for the other physicists whose work Einstein had studied and admired, such as Wilhelm Wien and Henri Poincaré, he surely must have hoped for some early and real understanding of what he was trying to do. But on that score, Einstein was to be completely disappointed; and Mach, after early expressions of brief, diplomatic, and cautious words of encouragement, turned against relativity (as we saw in Chapter 2) when he began to recognize what the program of relativity was and what it demanded. Hermann Minkowski's enthusiastic embrace in 1908 of relativity theory—in his own reinterpretation—left Einstein himself at first quite cold. For the next few years, younger scientists and philosophers, such as Friedrich Adler in Switzerland, Joseph Petzoldt in Germany, Paul Langevin in France, and Richard C. Tolman and Gilbert N. Lewis in the United States, began to adopt relativity for their own purposes. But again, more often than not, they initially misunderstood the main point. The same pattern of cases of either appropriation by misinterpretation or outright rejection continued in some circles for decades.

These various responses to a theory that now seems so clear to scientists call for explanation. In such matters one does not expect to find just one or two mechanisms, and not all of them need have been clear to the participants themselves. But even a brief list must contain facts such as these: that Einstein's first paper on relativity theory had even greater ambitions than those openly stated; that it was complex and strangely construed, as seen by those habituated in the then current style of physics—in effect a violation of the contemporary Rhetoric of Assertion in physics—whereas for us, inheritors of much of Einstein's way of thinking and arguing, the paper makes far fewer demands; that Einstein's proposals were really not *necessary* for a physicist in 1905 because what William James would have called the theory's "cash value" for contemporaries did not seem to be superior to those derived from, say, Lorentz's quite differently based and quite successful theory; that it asked for large conceptual sacrifices to be made (such as abandoning the absolutes of time and simultaneity, and the ether) in return for the relief from major pains that only the unknown young author seemed to feel.

To top it off, the paper in its published form, written hastily after years of reflection within five or six busy weeks, had—in addition to errors that soon had to be corrected[29]—a cavalier air about it. That is indicated, for example, by its unusual failure to have any bibliographic references and by its resistance to demonstrate clearly some of its own favorable points and implications, such as that what are still called the Lorentz transformation equations were now derivable very simply from Einstein's postulates and thus did not have to be introduced in a manner both Lorentz and Einstein considered ad hoc. (The simplest derivation had to be pointed out in a footnote, added by the editor when the paper was reprinted.) Recalling Aristotle's three kinds of "modes of persuasion" necessary for the good rhetorician—exhibiting the good personal character of the speaker, putting the audience in the right frame of mind, and providing a proof through the speech itself—we note that none of these seemed to weigh on Einstein. If anything, he seemed to be paying as little attention to them as possible.

I do not know which of these or other "flaws" were on Einstein's mind when he himself in the 1940s came to express displeasure with his 1905 paper. The occasion was the following, as related to me by his long-time secretary, Helen Dukas: Einstein had been asked to donate the manuscript of his 1905 paper to a fundraising drive on behalf of United States government war bonds. Because he had not kept that manuscript, he decided to make a new, handwritten copy from the published version. (It actually fetched a huge sum for the government in the auction and now resides in the Library of Congress.) To speed the work of copying, Einstein had Helen Dukas dictate the paper to him. She told me that instead of following her dictation faithfully, he repeatedly objected that he "could have said it much better," and indeed now intended to do so. She had to plead with him constantly to keep him from improving on his old work.[30]

At any rate, if one analyzes the 1905 relativity paper with care, line by line,[31] one can again discern throughout the existence of two Actors engaged in their different monologues, one with his past, the other with his future. A look at the first few lines will suffice here to make the point. The first paragraph is centered on a retrospective reflection of Actor 1 upon Einstein's early struggle with classical electrodynamics, as he experienced it in his student years, for example in reading August Föppl's text *Einführung in die Maxwellsche Theorie der Elektrizität,* 1894 (which in turn had acknowledged epistemological debts, particularly to Kirchhoff, Hertz, and Mach). The construction of Einstein's initial *Problemstellung* in the 1905 paper is completely parallel to Föppl's fifth main section, and includes especially the Faraday experiment referred to by both. The latter played, as Einstein repeatedly noted, "a leading role" in "the construction of the special relativity theory."[32]

In the second paragraph, we continue to hear echoes of the concerns of Einstein's earlier self, including the *Gedanken*-experiment at about age sixteen and the abortive plans for actual experiments, made while a student at the university. But we also begin to discern Actor 2 in the decision to remove the barriers separating the laws of physics, starting with those between mechanics and electrodynamics. For it was the most enduring passion of Einstein, from his earliest years as a scientist to the end, to pursue what he called (in a letter to W. deSitter) "my need to generalize" ("mein Verallgemeinerungsbedürfnis").

That need appeared already while he composed his first published paper (1901) on the unlikely subject of capillarity. He writes to his friend Marcel Grossmann (letter of 15 April 1901) that he is trying there to bridge the molecular forces and Newtonian forces at a distance, and he bursts out: "It is a magnificent feeling to recognize the unity [*Einheitlichkeit*] of a complex of phenomena which to direct observation appear to be quite separate things." Similarly, he reported in a manuscript written about 1920 that in writing the 1905 paper he had found the contemporary interpretation of the Faraday experiment "unbearable" because it regarded as "two fundamentally different cases" what he felt needed to be subsumed under one more general case.[33] And in virtually each of the other papers preceding *or* following upon the relativity paper of 1905, we find that the appeal of generalizing takes over and becomes a directive for research. We know now that even while working on special relativity Einstein felt it to be too limited, and hence decided to extend the postulate of relativity to nonuniformly moving coordinate systems.

To put it more starkly: When Einstein begins his work, he is aware that physicists are deeply divided between the program and claims of the mechanistic world picture and the electromagnetic world picture. Already in his third paper (written in 1902) on extending Boltzmann's ideas in thermodynamics and statistical mechanics, he joins the battle head-on by testing some limits of what he calls there the "mechanische Weltbild." By the time he is writing the relativity paper, he has seen that neither the mechanistic nor the electromagnetic world picture by itself suffices, for example, in dealing simply with fluctuation phenomena. Nor would a victory for one or the other have satisfied him; as he said later, it would, for example, leave us with "two types of conceptual elements, on the one hand material points with forces between them, and, on the other hand, the continuous fields, . . . an intermediate state of physics without a uniform basis for the entirety."[34] Without their having some awareness of these agendas of Actor 2 for the future, Einstein's paper must have been far more puzzling to his contemporaries than it is to us who know how it all turned out.

The motivating words "Weltbild" or "Verallgemeinerung" or "uniform basis for the entirety," of course, do not appear anywhere in the 1905 paper. But my point here is emphatically that just as there is a danger of blindly reading ahistorical elements back into earlier work, there is equally a danger to being blind to the forward thrust that may silently underlie the program of research at a particular time. One will not understand Actor 1, speaking at time *t*, without having made a detailed historical study of what preceded *t*. But one will not even properly hear Actor 2 unless one has studied what followed after *t*. Many excessively "internalistic" studies of a scientific publication have failed to catch the spirit of the work for that reason.

The Stage Fills

Now we turn our gaze more directly on the group of Actors who in our metaphor constitute the dramatic personae embedded in the text, although the scientist-author usually will claim to be giving us merely access to nature itself as revealed directly through "Reason and Experiments." In the case of Bohr's paper, we saw him turning to Rutherford, Thomson, and Planck by summarizing what he perceived to be correct and important or incorrect and incomplete about their prior work in this field. We could have added others; for example, Bohr has a lively though one-sided "conversation" with J. W. Nicholson (on pp. 6–7, 15, 23–24 of Part I of his paper, and more in Part II) about Nicholson's doomed theory of line emission spectra.

For the purpose of such "conversations," those other scientist-colleagues in the case are brought on the stage in the author's script explicitly or implicitly. But of course they are presented to us on the author's terms—their voices and proposals are adjusted or interpreted to serve the script. While Bohr was meticulously fair, and surely no distortion was intended, occasionally we do hear later from one of them in their own voice, when they take exception to what they perceive to have been a misunderstanding of their true position. At any rate, the Rutherford, Thomson, Planck, or Nicholson of whom we learn in Bohr's paper cannot, with the best will in the world, be considered fully representative of the originals. On this stage, alongside the two "Bohrs," they are Actors 3, 4, 5, 6 . . ., speaking lines that their corresponding models might not have thought of.

The same considerations apply to the main stage filled by the characters in Einstein's paper. There we encounter, in addition to Einstein serving as both Actors 1 and 2, also H. A. Lorentz—but only a mere fragment of the Lorentz we know to have existed in 1905, for Einstein had not yet read Lorentz's key paper of 1904,[35] and he was not convinced by those publications that he had read. We have already met Einstein's Föppl, one of several characters not mentioned by name in Einstein's paper. Ernst Mach is also not named; but a facsimile of him presides so visibly over the section "Kinematical Part" of Einstein's paper that a whole generation of positivistically inclined scientists and philosophers (from Petzoldt to Heisenberg) was misled to think of the whole paper as primarily a triumph of positivism. Other partially recognizable but anonymous Actors who make appearances bear fainter likenesses—

Helmholtz, Hertz, Boltzmann, Wien, Abraham, and, the faintest voice of all, David Hume. There may be others. For example, because we lack here the wealth of drafts and letters that we have for Bohr and his circle, written during the crucial period of composing his paper, we do not know which passages in Einstein's paper may refer directly to his conversations with friends, such as Besso.[36]

Einstein's representation of "Lorentz" is a particularly interesting character in his own right, as indeed is Lorentz's "Einstein" in Lorentz's later publications. After their first meeting in 1911, Einstein came to admire and even love Lorentz as a superb physicist and a remarkable person; and Lorentz's fondness for Einstein was also very deep when they came to know each other. But just as Lorentz never accepted relativity fully, Einstein had not much patience with Lorentz's approach to electrodynamics.

It is therefore highly ironic and appropriate for a study of rhetoric in science that during the early years the very different research programs of both men were widely subsumed in the literature under the joint name "Lorentz-Einstein." That fiction is worth more than a brief glance. One of the first to use it was Walter Kaufmann in early 1906, in the first article in the *Annalen der Physik* to respond to Einstein's 1905 paper—by putting "the Lorentz-Einsteinian fundamental assumption" to the test.[37] As we shall see in more detail below, Planck thereupon took up the cudgel on behalf of Einstein's relativity. But he too began his talk[38] with the line that "recently H. A. Lorentz, and in more generalized form Einstein, [had] introduced the Principle of Relativity"; and soon thereafter,[39] Planck, too, used the term "Lorentz-Einsteinian" theory.

Of course, there is a sense in which one may read Lorentz's 1904 paper and Einstein's 1905 work as operationally "equivalent"—another potent term that will deserve more than passing comment. Both theories used almost the same transformation equations and thus allowed effectively the same observable results to be derived with respect to experiments of interest at the time. Apart from that, however, the two theories were at opposite poles in every respect—in terms of their genesis, their physical and philosophical underpinnings, their respective assumptions, including the thematic ones, and their implicit further goals. In short, they were the products of quite different world views.

For example, as Lorentz's book of 1895[40] and the structure of his 1904 papers show, his work was driven largely by the strategy of patching up a theory of the electron that had been battered by puzzling recent experiments. On the other hand, Einstein's paper was, as he stressed over and over again, motivated by the desire to build a coherent physics "by the discovery of a universal formal principle" on the model of thermodynamics[41] and helped by his reinterpretation of old and well-known first-order experiments (Faraday's stellar aberration and Fizeau's measurements of light propagation in moving water ["They were enough," as Einstein told R. S. Shankland]). Lorentz did not hesitate to continue to introduce what he himself regarded as "somewhat artificial," ad hoc auxiliary conceptions as needed,[42] even after being scolded for it by Poincaré. And he freely confessed in 1912 that his theory of 1904, built around the model of a deformable, mechanically unstable electron, exhibited "clumsiness" and incompleteness, while only Einstein's provided "a general, strictly and exactly valid law."[43] Moreover, Lorentz's was essentially a physics of a particle, the electron, whereas Einstein's was a physics of any event in space and time. The ether provided obviously yet another demarcation criterion between the two, Lorentz's physics being firmly based on it to the end, while Einstein had dismissed it in an early passage with a casual wave of the hand.

We therefore do not find it surprising that their respective world pictures are entirely different also: on Lorentz's side, the best representation of the electromagnetic *Weltbild* available at the time; on Einstein's side, a new one that demanded applicability across all fields of physics, as well as the elimination wherever possible of asymmetries, ad hoc hypotheses, and redundancies (the existence of which Einstein found "unbearable"), no matter what cost it would entail in terms of resulting conceptual rearrangement. But such basic differences in the underlying world pictures were slow to be recognized,[44] and the long persistence of the term "Lorentz-Einstein" was an indicator of it.

An Experiment in the Rhetoric of Appropriation / Rejection

Having watched the main stage fill with agents implied in the internal, rhetorical space of Einstein's own paper of 1905, we can now visit an external *side stage,* on which the "real" versions of the characters carried out their acts of appropriation or rejection of what Einstein had to offer, immediately after his publication.

Here we are fortunate in that there took place a public encounter of opposites that may serve as an "experiment" in the Rhetoric of Appropriation / Rejection and reveals some of its fine structure. It was in fact initiated by the publication of Walter Kaufmann's papers of 1905 and 1906, claiming to give the empirical test data that would crucially decide between the current theories.[45] For our purposes we need only cite the results that this distinguished, Göttingen-based physicist himself put near the start of this major experimental examination (finished on 1 January 1906) of Einstein's 1905 work. Kaufmann wrote in italics: "I anticipate right here that the . . . measurement results are not compatible with the Lorentz-Einstein fundamental assumptions." Further on, again in italics, he declared those assumptions "a failure." For good measure, Kaufmann pronounced his data to favor a recent, much more limited theory by Max Abraham.[46]

Moreover, in an addendum less than four months later, Kaufmann implied that if one wanted to distinguish between these two discredited approaches, despite the equivalence of their derivable predictions of empirical facts, Lorentz's had one advantage over Einstein's. For the inductivist methodology of Lorentz had yielded in his case the proposed "independence of all observed phenomena from a uniform translation" as an "end result," whereas Einstein had merely proposed it initially "as a postulate, at the apex," achieving thereby the same system of equations "through pure mathematics."[47]

That was little comfort for Lorentz. Most respectful as always of the work of experimenters, this great theoretician saw his labors of well over a decade suddenly destroyed even by Kaufmann's preliminary (1905) results. He seemed devastated, writing to his friend and fellow theoretician Poincaré on 8 March 1906: "Unfortunately my hypothesis . . . is in contradiction with Kaufmann's results, and I must abandon it. I am thus at the end of my Latin." He appealed to Poincaré for help. But none came; instead, Poincaré noted that the "entire theory" may well be threatened by Kaufmann's results.[48]

Einstein's response to Kaufmann has also been noted before; it was completely different, not least perhaps because Einstein had—along with his long-standing interest in the experimental side of physics—a healthy skepticism about the latest news from the laboratory if its claims implied a modification of his closely-reasoned theories. At first, Einstein ignored the results, and it took an appeal from Johannes Stark in 1907 to survey the state of the relativity theory to move him. In brief, Einstein indicated that his theory had been taken too narrowly, as only a contribution to electrodynamics; that it was possible that the data from such a difficult experiment as Kaufmann's could be in sufficient agreement with his own theory after all; and that systematic errors in Kaufmann's data seemed likely.

But above all, Einstein's intuition told him something to which others were not alert: the data which seemed so conclusive may have been faulty *because* they favored theories, such as Abraham's and Bucherer's, which applied to a rather small region of physics compared with Einstein's own:

> In my opinion both theories have a rather small probability, because their fundamental assumptions concerning the mass of moving electrons are not explainable in terms of theoretical systems which embrace a greater complex of phenomena.[49]

In the meantime, there had recently taken place a most revealing debate concerning the grounds for believing in any of the theories in the absence of incontrovertible empirical evidence. It began with Max Planck's quick response to Kaufmann's publication. At forty-eight years of age one of the most distinguished physicists in the world, and well on his way to becoming the dour dean of German physics, Planck showed that he was sensitive to the deepest meaning of Einstein's 1905 paper. In a brief talk of 23 March 1906, he declared that if the "Principle of Relativity" (as he called the theory at first) were "borne out, it will be a grand simplification of all problems in the electrodynamics of moving bodies." He added that a thought of such "simplicity and generality" deserved, even in the face of Kaufmann's claimed disproof, to be subjected to more than just one test; and if the idea then did turn out to have been defective, it should nevertheless be taken *ad absurdum*, and its consequences examined.[50]

A few months later, Planck undertook a long, detailed reexamination of Kaufmann's recent result of experiments on the deflection of beta rays, which "for different electrodynamic theories is so-to-speak a question of life or death."[51] He then recast the theoretical base of Kaufmann's experiment more thoroughly than Kaufmann himself had done. While also revealing the considerable number of assumptions that Kaufmann needed (e.g., field homogeneity), Planck compared the reported observations with the expected values that can be calculated on the basis of "those two theories which so far have been most developed," that of Max Abraham (1903) "and the Lorentz-Einsteinian [as noted, Planck also made use of the term] in which the Principle of Relativity has full validity."

Even in this reexamination, Planck found the "data" that Kaufmann had published to be closer to the prediction of Abraham's theory than to the "Lorentz-Einsteinian"; typically, observation yielded the value 0.0247 (in the units Kaufmann used), while the first theory predicted 0.0262 and the second 0.0273. But as if to show that "data," too, have rhetorical uses, Planck does not see in those numbers "a definite proof of the first and a disproof of the second theory." After all, the differences between the predictions from the two theories were generally smaller than the difference between Kaufmann's reported "observations" and either of the theoretical values. Hence, Planck notes, one can begin to suspect a systematic error in the experiment or in its assumptions. "There seems to be a significant defect [*Lücke*]" somewhere; hence a definite decision between the theories is at this point unwarranted. Moreover, Planck finds the whole experiment a bit misguided, for it uses fast beta rays, whereas a better decision between the theories can be shown on theoretical grounds to be expected from the use of slower electron beams.

Happily, Planck had chosen to present these findings in a public lecture (Stuttgart, *Deutsche Naturforscherversammlung,* 19 September 1906), and after it there ensued a lively discussion that was also published.[52] For a study of the Rhetoric of Appropriation / Rejection, it is a revealing and even amusing theatrical script of its own. Walter Kaufmann rose first; he was glad to see that Planck's calculations, made on a different basis, had resulted in "identical numerical results"—only a slight exaggeration—and so gave one confidence that no errors of calculation had entered. But after all, Kaufmann had to insist, the Lorentz-Einstein (L-E) theory predictions deviated from his data throughout by 10 to 12 percent, whereas Abraham's (A) theory came to within 3 to 5 percent—also outside the error of observation, but possibly within the error from *all* sources.

Planck's response was uncharacteristically curt. In the absence of a full understanding of the error sources in addition to observation errors, it was for him quite thinkable that when such corrections eventually might be made, they would bring the data closer to the L-E theory than to its rival. A. H. Bucherer now rose to reflect in a rambling speech on how Planck's analysis affected his own theory, one similar to Abraham's, and how it might be improved. (In passing, he did make, as had rarely been done so far, a distinction between Lorentz's and Einstein's theories, both of which he believed to be flawed for different reasons; and he was the first person to adopt Planck's newly proposed term, *Relativtheorie,* but shortly after coined the term *Relativitätstheorie.*) However, for his labors he was rebuffed by Planck, who asked him about a "very important" test of Bucherer's theory, which Bucherer had to confess he had not yet made.

Now it was Max Abraham's turn to speak. It must be remembered that he was a brilliant physicist, whom Einstein also respected, but whose theory of the rigid electron was based on a completely different, fiercely held world picture. As von Laue and Max Born noted some years later, Abraham

> found the abstractions of Einstein disgusting in his very heart. He loved his absolute ether, his field equations, his rigid electron, as a youth loves his first passion whose memory cannot be erased by any later experience . . . [Einstein's] plan was to him thoroughly unsympathetic.[53]

Abraham rose and asserted (to "great laughter") that since the predictions from the L-E theory deviated from Kaufmann's data twice as much as his own theory, it followed that his own is "twice as good as . . . the *Relativtheorie.*" He was satisfied with the result. Moreover, his theory had the advantage that it was a "purely electromagnetic one." Even Lorentz's failed by that criterion because it assumed (as Poincaré had also found recently) the need for a term in addition to its electromagnetic energy.

Planck replied that he agreed with that fully—but so far Abraham's purely *elektrische* theory was only a hopeful postulate, an unachieved program. To be sure, the L-E theory "is also based on a postulate, namely that no absolute translatory motion can be discovered." So in Planck's opinion, we had here two unproved and undisproved theories. And at that crucial point, having put before his distinguished audience the choice between the two antithetical postulates, and hence between these antithetical conceptions of reality, the magisterial Planck added a few sentences that surely deserve to be a highlight of any future theory of the rhetoric of science:

> These two Postulates, it seems, cannot be united; and so it comes to this: to which Postulate [L-E or A] to give preference. *As to myself, the Lorentzian is really more congenial. [Mir is das Lorentzsche eigentlich sympathischer.]*

When the chips were down, the inner motivation for making a choice in the absence of meaningful differences obtainable through Newton's "Reason and Experiments" became visible: it is the *feeling of sympathy, of congeniality with one world picture rather than with its opposite, a decision based on one's scientific taste.*

Having made this revelation (in which he had condensed the term for the L-E theory further, into merely Lorentz's), Planck immediately added a protective sentence: "It would be best if both fields were to be further developed, and in the end experiment provided the decision." Thereupon Arnold Sommerfeld, at thirty-eight years of age one of the bright newer stars of physics, felt compelled to remark he could not join Planck in the "pessimistic point of view" that the decision should be delayed until experiments spoke more clearly. Leaving aside Kaufmann's results because the "extraordinary difficulties of the measurements" might well have produced deviations from the expected data that came from still unknown sources of error, Sommerfeld could make known his choice now:

> I suspect that regarding the question of Principles which Herr Planck has formulated, preference is given to the electrodynamic Postulate [i.e., the Abraham theory based on the electromagnetic world picture] by those under 40 years of age, and to the mechanistic-relativistic one [i.e., to the Einsteinian extension of the principle of relativity to all of physics] by those over 40 years. I prefer the electromagnetic one myself [laughter].

While Sommerfeld's division was not quite correct, and he soon changed allegiances, his confession also underlined that congeniality of point of view is a quasi-aesthetic criterion for theory choice in science—even as rhetoricians from classical Greece on knew it to be in the three traditional areas of display: political, forensic, and ceremonial.

In an anticlimactic ending of the discussion, Kaufmann got up once more. He objected that "the epistemological worth" of the relativity postulate was small because it did not apply to systems other than inertial ones—a point that Einstein, of course, knew well, and that was propelling him toward the "generalization" in which he was to succeed shortly. Planck squashed Kaufmann in three sentences: Kaufmann had missed the main feature of the relativistic point of view—that what could not be observed in inertial systems by mechanical experiments should also be unobservable by electrodynamic ones.

Nothing had happened that *forced* anyone to change his mind, to abandon one theory together with the world picture on which it was based, and to favor adopting the opposite. Individual experimental results of a narrow sort continued to come in for some years and lent themselves to one cause or the other, depending on how robust one thought the underlying assumptions were. Thus for a few years, the scientific community found itself somewhat at a loss how to deal with two such differently based theories whose "cash values" were about the same—and both of which were still under the cloud of the Kaufmann experiment (which was not fully unmasked as defective until 1916). In another irony that must have amused Einstein, the best support for both theories for years was thought to be the fact that in their fundamentally different ways each "explained" the haunting failures to find ether drift effects.

By Way of Epilogue: The Inertia of Rhetoric

When *was* it over? When did Bohr's and Einstein's "art of persuasive argumentation" succeed respectively in converting their community to a new theory or way of seeing the world? The superiority and scope of Bohr's theory, plus the results of decisive experiments (e.g., Franck-Hertz, 1914), had made it soon irresistible. But in the case of relativity, there was no hope of having some crucial experiment decide quickly between the rival theories, with their bases in vastly different world pictures. What had to happen, as so often must, was a slow process by which more and more of the visible members of the scientific community learned to hear and understand the voices on the stage for which Einstein had written the script. For example, the perceptive and well-placed physicist Wilhelm Wien, with whom Einstein had begun a correspondence in 1899, had initially published his disagreement with relativity; but by 1909 he had become persuaded by it and its world picture essentially on aesthetic grounds. He wrote:

> What speaks for it most of all, however, is the inner consistency which makes it possible to lay a foundation having no self-contradictions, one that applies to the totality of physical appearances, although thereby the customary conceptions experience a transformation.[54]

We noted that the appearance of Max von Laue's textbook of 1911, entitled significantly still *Das Relativitätsprinzip*,[55] essentially dates the first solid indication of the victory of Einstein's world conception over Lorentz's (and Abraham's); but even then von Laue had to confess that

> a really experimental decision between the theory of Lorentz and the Relativity Theory is indeed not to be gained; and that the first of these nevertheless had receded into the background is chiefly due to the fact that, close as it comes to the Relativity

> Theory, yet it lacks the great simple universal principle, the possession of which lends the Relativity Theory from the start an imposing appearance.

Indeed, after Kaufmann,

> very significant experiments by Bucherer [1909] and E. Hupka [1910] seemed to speak in favor of the Relativity Theory, but opinion about their power of proof is still so divided that Relativity, from that side, has not yet received unquestionably reliable support.

Von Laue added that the wealth of different phenomena encompassed by relativity theory was so vast that it was a task of the highest order to achieve an explanation of all of these by the adoption of one point of view. Thus "it is no wonder that this task reaches deeply into our whole physical world picture [*Weltbild*] and touches on the epistemological foundations of science."

It took some years more for the special relativity theory to become truly a widely accepted part of physics. That had to wait for developments far from the scope of Einstein's 1905 paper itself—foremost among them experimental successes, such as the eclipse expedition of 1919 with its test of a prediction of the general theory of relativity and the use of relativistic calculations to explain the fine structure of spectral lines.

In the meantime, the interested public and indeed some physicists had to seek support for the relativity theory, particularly in the face of its challenging paradoxes and iconoclastic demands, chiefly in the apparent ease with which it explained A. A. Michelson's results—an experiment that had counted little if anything for Einstein himself, but which to this day profits from being pedagogically the easiest tool of persuasion (at least in the oversimplified versions found in textbooks). Thus it came about that Einstein's scientific *Weltbild* has been absorbed into the culture of science and beyond with the aid of a rhetoric that had little to do with its genesis.[56]

A decade after his first edition, von Laue published the fourth edition of his successful text, renamed straightforwardly *Die Relativitätstheorie* (1921). By that time, two years after the famous general relativity test of November 1919, most physicists had come to accept Einstein's special relativity over Lorentz's relativistic electrodynamics. Yet, there still was a tendency to confuse them in certain profound respects; and for this reason, von Laue felt compelled to end volume 1 of his 1921 book with a special selection, largely taken from his first edition, in which he patiently tried to set the matter straight once more.

The historical sequence of developments, he wrote, had produced the misunderstanding that relativity theory is more closely related to electrodynamics than to mechanics. The source of this misperception is that transformation equations were indeed first deduced from electrodynamic considerations (by the majestic Lorentz in various, successively better forms, over several years ending by 1905) and most physicists will have absorbed the equations and their original electrodynamic context early in their training. But in modern relativity theory, the equations applied equally to the phenomena in *all* fields of physics, including mechanics, even though in that field one usually did not need them to make predictions that are correct to within the error limits of most measurements in mechanics.

A related misunderstanding was, he said, that since all forces of physics are subject to Lorentz transformations, they all may have a common origin, namely, the electrodynamic forces to which Lorentz had first applied them. But that thought, too, was entirely unwarranted. On the contrary, the fact that the relativity principle can be applied equally to all forces hints not at a subordination of mechanics to electrodynamics, but at the "equal subordination of both under higher laws."

Although von Laue did not speculate further on the reasons behind the long-lingering confusions, we may point to two that have roots in rhetoric: the "momentum" of the old term "Lorentz transformations" and the long-term persistence of the implication of operational "equivalence" between Lorentz's and Einstein's theories. We are dealing here with what one might call the inertia of rhetoric.

The "equivalence" of two (or more) differently based theories occurs again and again and is one of the surprising facts of science. Famous cases include the equally powerful (for making useful predictions) schemata of Copernicus and his Ptolemaic opponents, and the consequences derivable from either Heisenberg's matrix mechanics or Schrödinger's wave mechanics. Richard Feynman has put the puzzle in perspective in a memorable passage on yet another such case, that of the law of gravitation:

> Mathematically each of the three different formulations, Newton's law [of gravitation], the local field method, and the minimum principle, gives exactly the same consequences. What do we do then? You will read in all the books that we cannot decide scientifically on one or the other. That is true. They are equivalent scientifically. It is impossible to make a decision between them if all the consequences are the same. But psychologically they are very different, in two ways. First, philosophically you like them or you do not like them; and training is the only way to beat that disease. Second, psychologically they are very different because they are completely unequivalent when you are trying to guess new laws.[58]

"Guessing new laws" is here shorthand for getting at new science, advancing beyond the stage reached by the different formulations that yielded "equivalent" results on previous puzzles. But—more than in most other endeavors—getting at new science tomorrow is the main purpose of doing science today. So it is a matter of crucial significance that when the two theories are extrapolated beyond the intersection point where, for the needs of the moment, the predictions are (more or less) the same, the next steps on the diverging trajectories are going to be quite different. As we saw in the confrontation between Bohr and his critics, and in the comparison of the Lorentzian and Einsteinian theories, every major theory in science is shaped and propelled by its own list of themata and its own world view. Thereby each sets the stage for a future form of science quite different from its rival—a future stage on which a new cast of characters can present its own acts in the unending play.

Notes

1. Aristotle, *Rhetorica*, in *The Works of Aristotle*, vol. XI, ed. W. D. Ross (Oxford: Clarendon Press, 1924), p. 1356a of Book I.2.

2. Isaac Newton, *Opticks* [1730 edition] (New York: Dover Publications, Inc.), p. 1.

3. Ibid., p. 376. The classic position is summarized in Einstein's sentence, "The belief in an external world independent of the perceiving sub-

ject is the basis of all natural science"; *Ideas and Opinions* (New York: Dell Publishing Co., 1954), p. 260.

4. Aristotle, *Rhetorica,* pp. 1357b, 1359a.

5. For example, Chapter 2 of Gerald Holton, *The Scientific Imagination: Case Studies* (Cambridge: Cambridge University Press, 1978), and Chapter 8 of Holton, *The Advancement of Science, and Its Burdens* (Cambridge: Cambridge University Press, 1986).

6. Even this brief announcement of my main theme cannot be allowed to pass without a bow to a justly famous work based on analogous (but only analogous) suppositions. I refer, of course, to Alexandre Koyré's *Galilean Studies,* in which he demonstrated that the rhetorical repertoire of Galileo's *Two New Sciences* was divided among five agents / Actors: Salviati, Simplicio, and Sagredo, with their different voices and goals, on center stage throughout; but also, in apparently only brief appearances, the "Author" (Galileo himself) and, most important, the reader / auditor beyond the stage, over whose soul the four others tangle. And then we discover a sixth agent involved in the proceedings—the translators (for example, in the Crew-de Salvio version), who took the liberty of effectively falsifying the text to fit their particular pre-Koyréan epistemology.

The types of Actors involved and the division of labor between them will be rather different in the cases to be considered here. Still, Koyré's type of analysis has been shown to have relevance even in understanding discussions in contemporary science, e.g., in the study of the unintentionally tape-recorded interactions among three astronomers as they discovered the first optical pulsar, at Steward Observatory on 16 January 1969.

7. I shall assume here, rather than repeat or summarize, the elements of the thematic analysis of scientific thought.

8. Michel Foucault, *The Order of Things* (New York: Random House, 1973), p. 159. See also p. xi.

9. Emilio Segrè, *From X-Rays to Quarks* (New York: W. H. Freeman and Co., 1980), p. 127.

10. Leon Rosenfeld, ed. *Niels Bohr, On the Constitution of Atoms and Molecules* (New York: W. A. Benjamin Inc.), p. xli.

11. Abraham Pais, *Inward Bound* (Oxford: Clarendon Press, 1986), pp. 208–211.

12. A significant but neglected study site is what scientists at the frontier regard as absurd, ugly, unbearable. For some of Einstein's explicit statements on what he regarded as "unbearable" in scientific theory, and on the aesthetic elements characterizing good theory generally, see chapters 2 and 4 in Holton, *The Advancement of Science* (cit. n. 5).

13. *Nature* (November 6, 1913), 92: 2297, p. 306.

14. Niels Bohr, interview of 7 November 1962, p. 1, in American Institute of Physics transcript of *Sources for the History of Quantum Physics* (SHQP).

15. Pais, *Inward Bound,* p. 196 (emphasis in original).

16. Rutherford's letter is reprinted in N. Bohr, *Proceedings, Physical Society, 78* (1961), 1083, and Bohr's letter in Rosenfeld, *Niels Bohr,* p. xxxviii.

17. J. H. Jeans, "Report on Radiation and the Quantum Theory" (London, *The Electrician,* 1914), p. 89. And not only in the "new mechanics" of the atom—during the first dozen years of the new century, the thema (and hence the rhetoric) of discontinuity had sprung up also in fields as separate as genetics and radioactivity (including "mutation," "transmutation").

18. Ibid., translated from Henri Poincaré, "L'Hypotheses des Quanta," in *Dernières Pensées* (Paris: Flammarion, 1913), p. 90.

19. Wassily Kandinsky's autobiographical sketch about the years 1901–1913, in his book *Rückblick* (Baden-Baden: Woldemar Klein Verlag, 1955), p. 16, indicates how he overcame a block in his artistic work at that time: "A scientific event removed the most important obstacle: the further division of the atom. The collapse of the atom model was equivalent, in my soul, to the collapse of the whole world. Suddenly the thickest walls fell. I would not have been amazed if a stone appeared before my eye in the air, melted, and became invisible. Science seemed to me destroyed . . ."

20. Bohr interview, SHQP, 31 October 1962.

21. Ibid.

22. Ibid.

23. Quoted in Rosenfeld, *Niels Bohr,* p. xxii.

24. Holton, *Thematic Origins of Scientific Thought* (Cambridge, Mass.: Harvard University Press, revised edition, 1988), p. 156.

25. Quoted in Rosenfeld, *Niels Bohr,* p. xxxiv. Compare Alexander von Humboldt, *Cosmos: A Sketch of a Physical Description of the Universe,* trans. E. C. Otte (London, 1848), vol. I, p. 68: "The charm that exercises the most powerful influence on the mind is derived less from a knowledge of that which is, than from a perception of that which *will be.*"

26. ". . . this constant [*h*] is of such dimension and magnitude that it, together with the mass and charge of the particles [electrons], can determine a length of the order of magnitude required [i.e., for the atom]." Bohr does not even stop to write it down: Radius of the atom $\approx h^2/me^2$.

27. Cf., "The Resistance to 'Reckless' Hypotheses" in Holton, *Thematic Origins,* pp. 164–169.

28. Ibid., p. 246.

29. A significant infelicity in the definition of force was pointed out shortly thereafter by Max Planck. And in the collection of Einstein's own reprints from his desk, given to me by the estate after helping it to organize the archives, there appeared in Einstein's handwriting on the reprint of the 1905 paper a number of corrections to the printed version.

For documentation of the steps and timetable of Einstein's composition, see John Stachel, ed., *The Collected Papers of Albert Einstein,* vol. 2 (Princeton: Princeton University Press, 1989), pp. 253–274, especially pp. 261–266.

30. A priceless opportunity was thus missed. One can speculate, for example, that in a "revision" Einstein would have shown more clearly the close subterranean connections that have been shown to exist between the relativity paper and the earlier, so-different-appearing papers of 1905, on light emission and Brownian motion. One can also see that the internal structure of the three papers is parallel.

31. As has been done for other purposes in the monograph by Arthur I. Miller, *Albert Einstein's Special Theory of Relativity* (Reading, Mass.: Addison-Wesley Publishing Co., Inc., 1981). Among Miller's other contributions that are useful for the topic of this paper, see especially, "The Physics of Einstein's Relativity Paper . . .," *American Journal of Physics, 45* (1977), pp. 1040–1048, and "On Einstein's Invention of Special Relativity," in A. I. Miller, *Frontiers of Physics 1900–1911* (Boston: Birkhäuser, 1986), pp. 191–216.

32. Quoted in Holton, *Thematic Origins,* 1988, p. 381. Where not otherwise identified in what follows, brief passages or phrases quoted from Einstein will be found also in that source, Chapters 6–9.

33. Holton, *Thematic Origins,* 1988, p. 382. This early date contradicts a recent speculation that Einstein ascribed great importance to these experiments only when he reflected on the genesis of relativity in his old age.

34. Albert Einstein, "Autobiographical Notes," in *Albert Einstein, Philosopher-Scientist,* ed. P. A. Schlipp (Evanston, Ill.: The Library of Living Philosophers, 1949), p. 27. Miller, in *Frontiers,* p. 200, documents that by 1905 Einstein's "investigations of the structure of light revealed that classical electromagnetism failed in volumes of the order of the electron's. Thus, the electromagnetic worldpicture could not succeed. His Brownian motion investigations had yielded a similar result for mechanics; hence, exit any possibility for a mechanical worldpicture." Additional support comes from Einstein's letters, e.g., to M. von Laue and C. Seelig. In one of the early, neglected statements, a letter to E. Bovet of 7 June 1922 (*Wissen und Leben,* vol. 24 [1922], p. 902), Einstein refers to his theory as

"an improvement and modification of the basis of the physical-causal worldpicture."

35. H. A. Lorentz, "Electrodynamic Phenomena . . .," *Proceedings of the Royal Academy of Amsterdam,* 6 (1904), p. 809.

36. Miller, *Special Theory,* Section 1.15.1, has a useful discussion of the sources on which Einstein drew ("definitely," "very probably," "maybe").

37. Quoted in Holton, *Thematic Origins,* p. 253.

38. M. Planck, "Das Prinzip der Relativität . . .," *Verhandlungen, Deutsche Physikalische Gesellschaft,* 4 (1906): 136–141.

39. M. Planck, "Die Kaufmannsche Messung . . .," *Physikalische Zeitschrift,* 7 (1906): 753–761.

40. H. A. Lorentz, *Versuch einer Theorie* (Leiden: Brill, 1895).

41. Quoted in Holton, *Thematic Origins,* p. 310.

42. Lorentz, "Electromagnetic Phenomena." See also *Versuch einer Theorie,* in which Lorentz calls one of his own hypotheses initially "estranging" (p. 123) and comments on another one that "to be sure, there is no basis for it" (p. 124).

43. Quoted in Holton, *Thematic Origins,* p. 321.

44. The main oppositions between the two world pictures, showing at a glance how fundamentally they differ, are displayed graphically in William Berkson, *Fields of Force: The Development of a World View from Faraday to Einstein* (New York, 1974), p. 254.

45. These papers have been analyzed well by Miller, *Special Theory,* pp. 225–235, 333–352, and in his *Frontiers,* Essay 1.

46. W. Kaufmann, "Über die Konstitution . . .," *Annalen der Physik,* 19 (1906): 487–553.

47. W. Kaufmann, "Nachtrag . . .," *Annalen der Physik,* 20 (1906): 639–640.

48. For the Lorentz-Poincaré responses, see Miller, *Special Theory,* pp. 334–337.

49. A. Einstein, "*Relativitätsprinzip. . . .,*" *Jahrbuch der Radioaktivität und Elektronic,* 4 (1907): 411–462.

In a brief article dated August 1906 (A. Einstein, "Über eine Methode . . .," *Annalen der Physik,* 21 [1906]: 583–586), Einstein had mentioned Kaufmann's experiment with beta rays in passing, but studiously avoided discussing it, commenting on its results or worth, or even giving the bibliographic reference to it. Instead, Einstein proposed a different and better experiment to test the rival theories, one using slow cathode rays (apparently unaware that Planck had recently made a similar suggestion). Einstein predicted significant differences to result that would distinguish among three theories, namely, "Theorie von Bucherer," "Theorie von

Abraham," and "Theorie von Lorentz und Einstein." For the last, we note the use of the singular and of his own name in his text—all quite uncharacteristic for him. For this brief piece only, Einstein adopted or aped Kaufmann's terminology, and probably did so with tongue in cheek.

50. Planck, "Das Prinzip der Relativität."

51. Planck, "Die Kaufmannsche Messung."

52. *Physikalische Zeitschrift,* 7 (1906): 759–761.

53. Quoted in Holton, *The Scientific Imagination,* p. 10.

54. W. Wien, *Über Elektronen,* 2d ed. (Leipzig: Teubner, 1909), p. 32.

55. Braunschweig: F. Vieweg & Sohn, 1911. See especially pp. 18–21. (In 1911 the author's name was "Laue"; the *von* was added in 1914, when his father was made a member of the hereditary nobility.)

56. A whole study on the Rhetoric of Appropriation / Rejection could be devoted to the events subsequent to the publication in 1969 of the results of a historical study of the documents that showed we must take Einstein at his word that the genetic influence of the Michelson experiments was at best indirect and small. This finding contradicted the almost universal agreement at the time among historians and philosophers of science, as well as textbook writers, who held that the experiments had been a crucial guide for Einstein. That view was part of, and reinforced, a branch of the philosophy of science that incorporated experimenticism.

Since that publication in 1969 (see Holton, *Thematic Origins,* Chapter 8 and pp. 477–480), all additional, first-hand documents have supported the conclusion reached in 1969. Moreover, by and large it has been fully incorporated into the current view. But even though pedagogic texts now have to deprive themselves of the simplistic way to make plausible the origins and necessity of relativity, many of them still feel required to go through the motions, hoping thereby to persuade their students more easily about this demanding and counterintuitive theory. Thus the authors of *The Story of Physics,* Lloyd Motz and Jefferson Hane Weaver (New York: Plenum Press, 1989), pp. 252–253: "Physics [about 1900] was growing rapidly . . . but one experiment brought consternation and confusion—the famous Michelson-Morley experiment. Since this experiment had a direct bearing on the acceptance of the theory of relativity, *although historical evidence indicates that Einstein did not know about it when he wrote his relativity paper,* it is useful to examine this experiment" (italics supplied). And they proceed to do so at length.

57. Richard Feynman, *The Character of Physical Law* (Cambridge, Mass., and London: MIT Press, 1967), p. 53. I thank S. Sigurdsson for having drawn my attention to this passage.

4

On the Jeffersonian Research Program

The ambitious project of the Vienna Circle for an International Encyclopedia of Unified Science, discussed in Chapter 1, is a reminder of the historic debts modern science owes to the program of the enthusiastic encyclopedists of the Enlightenment. But the insufficiency of the penchant for accumulation to lead by itself to good science was captured well in Einstein's remark, interpreting Ernst Mach's limited goal as consisting "in a mere ordering of empirical material,"[1] and thus leading to nothing more than an encyclopedia of scientific phenomena—an activity reminiscent of the standard caricature of Babylonian science, the cumulation of data without integrating them into a network of theories. By contrast, the two main roots of modern science, reaching back to the seventeenth century, are usually and without too much distortion referred to as "basic science," identified with Newton, and "applied science," identified with Francis Bacon. But there is a third style of scientific praxis, one rarely recognized explicitly but deserving just now more attention than ever. The convenient term for it will be the "Jeffersonian research program." The reason for this terminology, and its appropriateness, will become clear when we look at Jefferson's activities in science.

At first glance, Jefferson's widely dispersed interests and his hunger for new phenomena seem to mark him to some degree as a follower of a Babylonian quest. But this will turn out to be a shallow estimate of his real aims. To clarify it, one must dispose first of the question whether Jefferson regarded himself as seriously involved in science of any style. History knows him first of all as President of the United States from 1801 to 1809. However, Jefferson was not comfortable with his characterization as political figure. He saw himself first of all as a student of the sciences, philosopher, educator, planter, and scholar, and even on his gravestone he did not permit any reference to be made to his presidency or to any of his other political achievements. Nor did he hold earlier political figures in highest esteem; rather it was of Francis Bacon, John Locke, and Isaac Newton that he wrote (letter of 15 February 1789 to John Trumbull), "I consider them as the three greatest men that have ever lived, without any exception."[2] On hearing of his election as Vice President, he wrote to James Madison (1 January 1797), "I am unable to decide in my own mind whether I had rather have it or not have it."[3] Coming to the nation's capital, then at Philadelphia, in early 1797, he was installed first as the newly elected President of the American Philosophical Society—calling this in the letter to the Secretaries of the Society "the most flattering incident of my life"[4]—and on the next day as Vice President of the United States. A few days later he returned to the American Philosophical Society to give a scientific paper, of which we shall hear more below. A passage in a later letter to John Adams catches his spirit: "I have given up newspapers in exchange for Tacitus and Thucydides, for Newton and Euclid, and I find myself much the happier."[5]

About the pursuit of scientific knowledge, Jefferson was never reluctant or regretful. In his early years, he came under the influence of Professor William Small at the College of William and Mary, under whom he studied mathematics and Newton's *Principia.* That contact, he said later (*Autobiography,* dated 6 January 1821), "probably fixed the destinies of my life."[6] His correspondence from beginning to end shows him never more relaxed and enthusiastic than when discussing scientific matters large or small. "Nature," he said, "intended me for the tranquil pursuit of science, by rendering them my supreme delight." His curiosity and persistence seemed infinite. Thus he kept his Garden Book for fifty-eight years continuously, noting when trees flowered, how experimental plantings fared, and so on. For years on end, he recorded the weather daily several times, even on the greatest day in American history, the fourth of July 1776—at 6 A.M., 9 A.M., 1 P.M., and 9 P.M. Jefferson therefore has kept historians of science busy writing articles about "Jefferson as a naturalist," "Jefferson as a vaccinator," "Thomas Jefferson as meteorologist," Jefferson on prehistoric Americans, on ethology, geography, botany, paleontology, mathematics, weights and measures, eugenics, agriculture, archeology, astronomy, medical theory and practice, and on and on.[7]

This brings us to the question whether his encyclopedic interests sufficiently characterized his approach to science. On the surface this would seem plausible, if only because of his insistent pursuit of encyclopedias of every kind. He owned several (among others, the *Chambers' Cyclopedia* of 1751–1752, the *New and Complete Dictionary of Arts and Sciences* of 1763–1764, the American reprint edition of the *Encyclopedia Britannica,* 3rd ed., 1788–1797, which itself contained several pages of text taken from Jefferson's *Notes on the State of Virginia;*[8] he wrote often to obtain copies for himself and for others (e.g., letter to William Short, 27 April 1790);[9] he advised James Madison to include the *Encyclopédie méthodique* on a "list of books proper for the use of Congress" (report of 24 January 1783);[10] in fact he ordered a copy of the *Encyclopédie méthodique* for public use in early 1781, but he apparently became so engrossed in it that in July 1782 a resolution had to be passed "to take measures for getting from Mr Jefferson the Encyclopaedia belonging to the public."[11]

Indeed, in Jefferson's collected papers there are well over a hundred references to various encyclopedias. It therefore was quite natural that Jefferson became, if only anonymously, a contributor to the new *Encyclopédie méthodique,* by his collaboration in 1786 with Jean Nicolas Démeunier on an article on the United States. He was so impressed by that encyclopedia that he took it upon himself to become its salesman: writing to James Monroe from Paris in June 1785,[12] at the end of a long letter on treaty negotiations, Jefferson offered to send Monroe the first forty volumes of the new edition. For good measure, a few months later, writing to James Madison, he offered to buy him a good set of the older edition. Moreover, Jefferson's *Notes on the State of Virginia,* written in 1781–1782, was truly an encyclopedic survey of his home state, with such headings as Boundaries, Rivers, Ports, Climates, Populations, Aborigines, Laws, Weights and Measures. Silvio Bedini correctly commented that it "received wide acclaim as the first comprehensive study of any part of the United States, and as one of the most important works derived from America to that time."[13]

Yet, Jefferson's encyclopedic attention and pursuits were not in the service of mere accretion; they were illuminated by the spirit of the Enlightenment, for he held that given enough scientific information, all human ills would yield to the serious application of human intelligence: "The ingenuity of man leaves us to despair of nothing within the laws of nature." It led him to dramatize and promote the sciences as no high public official in American history has done since, and sometimes at political risk to himself. His enemies seized on his interest in natural history: one said he should resign as President because surely he must be deranged, as proved by his passion for collecting and exhibiting new animals and searching for the bones of the extinct mastodon—a passion which also, as we shall see shortly, had a larger purpose.

The question of just how good Jefferson's science was has been often debated, for example in the fine study by John C. Greene, *American Science in the Age of Jefferson.*[14] Greene sees Jefferson as an inveterate connoisseur and promoter of science, but not as a serious scientist when measured against the leaders of science in Europe at that time—one thinks here of Laplace, Lavoisier, Thomas Young, Humphry Davy, and others. In such discussions an apologetic note is bound to enter. And to be sure, Jefferson did contribute much less directly than did Benjamin Franklin.

But in all this one must focus on the fact that Jefferson had a good understanding of the heart of the scientific method. Illustrations of this statement are not difficult to find. In early 1796, for example, Jefferson heard of the finding, in his own state of Virginia, of the fossilized remains of a huge-clawed animal. He obtained the bones and speculated with delight that they might be those of a hitherto unknown, monstrous American lion. He named it *Megalonyx* ("big claw"), although soon thereafter it was correctly identified as an extinct giant ground sloth. He studied the bones, arranged that they be sent to the American Philosophical Society, and wrote a memoir describing them. Thus it came about that on 10 March 1797, Jefferson, just sworn in as Vice President of the United States, had his paper read to the members of the Society while he presided over the meeting.

The essential point for us here is this: toward the end of his paper,[15] Jefferson took issue with the unfortunate idea, which was then current among French scientists, that the climate in America was so disadvantageous that it tended to produce stunted and degenerated life forms. Jefferson had long been annoyed by this slight to his country. He had tried to shake the French naturalist Buffon by sending him the skeleton of an enormous American elk, three times the bulk of the European kind. Now in his new paper, Jefferson took the opportunity again both to contribute to science and to answer Buffon's challenge. To Jefferson, the discovery of the bones of a giant animal was evidence against the European theory of degeneracy. This must have been very satisfying to him. Yet, at the end of the paper when Jefferson summed up the lesson of his new find, his fair and scientifically rational mind drew the right balance. He wrote:

> Are we then from all this to draw a conclusion, the reverse of that of Monsieur de Buffon. That nature has formed the larger animals of America, like its lakes, its rivers, and mountains, on a greater and prouder scale than in the other hemisphere? Not at all, we are to conclude that she has formed some things large and some things small, on both sides of the earth, for reasons which she has not enabled us to penetrate; and that we ought not to shut our eyes upon one half of her facts, and build systems on the other half.[16]

Jefferson was no less a patriot for being comfortable with scientific thinking. Again and again, he hoped to conquer superficial variety and differences by embracing themata fundamental to science, namely generalization and unification. In the Declaration of Independence, he had written that all men are created equal; now he was saying that on the whole all animals are created equal, regardless of location on the globe—just as elsewhere he wrote that the Indians of North America, save for the accident of opportunity and circumstance, were on a par with the rest of mankind; just as in Jefferson's bill of 1777 for the establishing of religious freedom, he provided for freedom of the profession of any religion whatever; and just as he dared to propose universal education, as one of the first to do so.

One can easily multiply the examples of Jefferson's sound scientific instinct. To be sure, by the standards of either Newton's science or Bacon's science, those previously mentioned, somewhat apologetic evaluations of Jefferson's science are defensible. But one aim of this chapter is to indicate that such evaluations draw attention away from a significant accomplishment: Jefferson's main contribution in this area was that he pointed toward a specific way of doing science, a model fundamentally different from the two standard ones against which he has habitually been measured—a third model, one that is still struggling to come to prominence in our time.

For consider the first of the two standard models, what might be called the "Newtonian research program," so named because Newton, while not the first to pursue it, was so explicit about this motivation. We recall that in the preface to the *Principia,* Newton described his aim and procedure: the observable phenomena (such as the fall of objects to earth, and some celestial motions) led him to postulate the existence of one general force of gravity by which all bodies attract one another, and from this in turn he was able to deduce in detail "the motion of the planets, the comets, the moon, and the sea."[17]

But no sooner had he acknowledged this stupendous achievement than he added, with an almost audible sigh of disappointment: "I wish we could derive the rest of the phenomena of Nature by the same kind of reasoning from mechanical principles."[18] All the other phenomena of Nature also! That is to say, optics, chemistry, the operation of the human senses . . . There is the Holy Grail: the mastery of the whole world of experience, by subsuming it ultimately under one unified theoretical structure. Einstein also taught that what he called the noblest aim of science was the attempt to grasp the totality of empirical facts, leaving out not a single datum of experience. Max Planck, too, asserted that the aim of science is "the complete [intellectual] mastery of the world of sensation."[19] It is not inappropriate to characterize this as the Newtonian research program. In short, it is the search for *omniscience.*

By contrast, the alternative vision for science may be called the "Baconian research program," for it was most eloquently defended by Francis Bacon and his followers. This style concentrates on science in the service of *omnipotence*, or as Bacon had put it, on "the enlarging of the bounds of Human Empire, to the effecting of all things possible."[20]

It is not our concern here to question just to what degree these shorthand labels are fully appropriate, and to what extent these two styles have on occasion interpenetrated each other in practice. Rather, we shall attend to what can be called the "Jeffersonian research program," a third mode, neither Newtonian nor Baconian, but—as befits the man who saw both Newton and Bacon among his chief heroes—one that takes something from each of these two programs and combines them for a new purpose. To put it briefly: *this style locates the center of research in an area of basic scientific ignorance that lies at the heart of a social problem.* It is therefore neither purely discipline-oriented nor purely problem-oriented (the latter being largely the application of *existing* basic knowledge for meeting supposed needs). The Jeffersonian type of research project, by contrast, is characterized by a combined mode, positioned intentionally in uncharted areas on the map of science itself, but it is motivated also by a credible perception that the findings may sooner or later have a bearing upon a persistent national or global problem. It is a combined mode that reflects the fact that Jefferson himself saw two intertwined goals for science—not only the fuller understanding of nature, but also what he called simply "the freedom and happiness of man."

In recent decades it has been realized more widely that many, perhaps most, of society's problems cannot be cured, or even properly understood, through existing technological, managerial, or political means alone. True solutions will often depend on making advances in basic science itself. Contrary to current folklore, it is not the progress of basic science that chiefly is to blame for many of the large-scale functional difficulties we face today; it is rather the *absence* of some specific fundamental scientific knowledge. This realization vastly enlarges the framework within which the scientist can look for research problems.

Examples readily come to mind. It is customary, for instance, to say that the population explosion is caused in part by the advance of applied medical science (better sanitation, diet, inoculation, antibiotics), many of which followed new developments in biology.

But one can claim equally well that the population explosion is bound to overwhelm us precisely because we do not yet have at hand sufficient knowledge in pure science. That is to say, the complex problem of overpopulation is due in a significant degree to our current (and long-acknowledged) ignorance of the basic process of conception—its biophysics, biochemistry, and physiology. No wonder that attempts at containing population growth have been so halting, and so easily defeated by ideological and political opposition.

The validity of the combined-mode positioning of research projects is well illustrated by referring to one of the chief episodes in Jefferson's life, when his thoughts turned to scientific matters, specifically his plans for the exploration of the North American continent, culminating eventually in his commissioning the Lewis and Clark expedition of 1803–1806.

On this expedition a great deal has been written, but not, it seems to me, with the proper recognition of Jefferson's combination of aims. To this day the enterprise is a superb example of organizing both the intellectual and the physical resources of a nation. And Jefferson himself pointed to his twin purpose for this expedition. He wrote that the aim was

> to extend for [the citizens of this nation] the boundaries of science, and to present to their knowledge that vast and fertile country which their sons are destined to fill with arts, with science, with freedom and happiness.[21]

To set the stage for a closer understanding of Jefferson's aim in launching the expedition, we must remember that there was intense interest, at least among the literati, even in earlier, colonial America, in all aspects of that vast, promising, and largely unknown continent, and on all subjects from topography to its native inhabitants. We see this in the charters and activities of such organizations as the American Philosophical Society and the American Academy of Arts and Sciences, which were founded before the achievement of statehood. Jefferson was a child of the frontier, the son of a surveyor and explorer who in 1751 had helped to make the first good map of Virginia. Even as a boy, Jefferson would have heard of plans to explore the continent. Indeed, to Jefferson, whose spectrum of curiosity was enormously wide, the continent—Jefferson called it "terra incognita"—was a grand museum, just waiting to be charted.

The sequence of Jefferson's early, failed attempts leading to the Lewis and Clark expedition is well known, for example the plans made with the adventurous John Ledyard in Paris in 1786. Even earlier, on 4 December 1783,[22] still as a private citizen, Jefferson proposed to General George Rogers Clark that he lead an expedition "for exploring the country from the Mississippi to California." Nothing came of that either. But the urgency of such a project was greatly increased in 1792 by the achievement of Captain Robert Gray of Boston, whose ship, the *Columbia Rediviva,* had been the first American vessel to circumnavigate the globe in 1787–1790. Gray reached the mouth of "Columbia's River" in Oregon on May 12. He entered it, went some thirty-six miles upstream, named the river, and thus set claim for the United States for sovereignty over the valley, over the watershed of the river, and over the adjacent coast, in accordance with a tradition widely respected internationally.

Gray's achievement did not become known in the east of the United States until the end of July 1793. By April of that year, Jefferson had been able to launch yet another attempt at a first expedition to explore what later came to be called the Territory of Louisiana, that vast and largely unmapped territory between the Mississippi River and the Rocky Mountains of the West, then all still owned by Spain. It was supposed to be a small exploring party. This time, Jefferson was planning it in his capacity as Vice President of the American Philosophical Society, arranging with the French botanist André Michaux to launch the transcontinental exploration. The budget was not to exceed four hundred dollars, to be raised by subscription.

It was very much an expedition of a learned society—and it also failed. But the most important component, which survived for later service, was Jefferson's document of instructions of April 1793 to the explorer Michaux.

> . . . the chief objects of your journey—Jefferson wrote—are to find the shortest and most convenient route of communication

> between the United States and the Pacific ocean, within the temperate latitudes, and to learn such particulars as can be obtained of the country through which it passes, its productions, inhabitants, and other interesting circumstances . . . You will, in the course of your journey, take notice of the country you pass through, its general face, soil, rivers, mountains, its productions animal, vegetable, and mineral so far as they may be new to us and may also be useful or very curious; the latitudes of places or materials for calculating it by such simple methods as your situation may admit you to practice; the names, numbers, and dwellings of the inhabitants, and such particularities as you can learn of their history, connection with each other, languages, manners, state of society and of the arts and commerce among them. Under the head of Animal history, that of the Mammoth is particularly recommended to your enquiries . . .[23]

We recognize here again the pen of the author of the *Notes on Virginia* of a dozen years earlier, including his preoccupation there with the fossil remains of the mammoth that had been found in Ohio, and the tradition among Indians, which Jefferson hopefully believed in, that a gigantic creature of this sort still existed in the northern part of the continent.

In 1801, eight years after Michaux's failure, all these pieces—from Jefferson's early, indiscriminate love for science, his fascination with the frontier, his encyclopedic study of his own state, the gaining by the young Republic of a foothold at the mouth of the Columbia River, and so on—all these pieces at last came together when Jefferson assumed the United States presidency. He had appointed as his private secretary Meriwether Lewis, who when not yet twenty had begged to be allowed to go on Michaux's ill-fated journey. The itch to explore the continent was in the blood of both men. But Jefferson now had a double rôle: he was the Chief of State of a vigorous young nation with a growing population, and he was also the man who confessed that the studies of science had fixed the destinies of his life and were his supreme delight.

By the end of 1802, before the great opportunity of the Louisiana Purchase offered itself, Jefferson wrote to the British and Spanish Ministers in Washington to find out what their government's reaction would be if a party of explorers were sent up the Missouri

River and across the mountains to the Pacific. The British Minister informed his government as follows:

> The President has for some years past had it in view to set on foot an expedition entirely of a scientific nature for exploring the Western Continent of America by the route of the Great River Missouri . . . He supposes this to be most natural and direct water-communication between the two Oceans, and he is ambitious in his character of a man of letters and of science, of distinguishing his Presidency by a discovery, now the only one left to his enterprise . . . [24]

Greene adds: "Such indeed was Jefferson's purpose, although he was careful to express the commercial, military and diplomatic benefits that would accrue from the expedition in his message to Congress requesting funds for the enterprise."[25]

On this view, Jefferson's aim was indeed "an expedition entirely of a scientific nature." But others have strongly urged a very different view. In his essay, "The Purpose of the Lewis and Clark Expedition," Ralph B. Guinness called it a "politico-commercial" venture, for which fur trade was "the primary concern,"[26] and Bernard DeVoto, in his book significantly entitled *The Course of Empire,* characterized the expedition as "an act of imperial policy."[27] The debate here is between those who see the exploration essentially in the service of the Newtonian program, in the pursuit of omniscience, *versus* those who see it as a Baconian program of enlarging the bounds of human powers, part of the search for omnipotence. But here, as so often, Jefferson was neither destined nor content to follow the models of others. He was capable of his own originality, of seeing the possibility of a marriage between the Newtonian and Baconian programs.

In April 1803, to everyone's surprise, Napoleon offered to sell the Americans the whole Louisiana Territory, chiefly to obtain resources for his war on Great Britain. The transaction went quickly, and almost doubled the territory of the United States. But it is very significant that three months before Napoleon's offer, Jefferson had already taken advantage of the expiration of an act for establishing trading houses with Indians in border areas, to ask Congress, in a confidential address, for authority and funds for yet another of his attempts at cross-continental exploration. On 18 January 1803 he proposed that "an intelligent officer, with ten or twelve chosen men . . . might explore the whole line, even to the Western ocean."[28] He assured Congress that the nation claiming sovereignty over those territories would regard this "as a literary pursuit, which it is in the habit of permitting within its dominions." He asked Congress for twenty-five hundred dollars and to label the venture so as to "cover the undertaking from notice" as an attempt "of extending the external commerce of the United States."[29]

Congress complied on 28 February 1803; but again, Jefferson had already appointed Lewis to lead the expedition, with William Clark, and had begun to arrange with scientific friends that Lewis be given instructions by them. For example, writing on 27 February 1803 to Benjamin S. Barton of the American Philosophical Society, Jefferson asked him to tutor Lewis to identify quickly new objects "in the lines of botany, zoology, or of Indian history." He adds, "I make no apology for this trouble, because I know that the same wish to promote science which has induced me to bring forward this proposition, will induce you to aid in promoting it."[30]

Jefferson's instructions to what became known as the Lewis and Clark expedition overlap substantially with those he had given years earlier to Michaux. As Jefferson spells it out in great detail over many pages (draft of 1803, final letter to Lewis signed on 20 June 1803), it is the dream of a naturalist, encyclopedist, and surveyor, merged with the vision of the statesman:

> The object of your mission is to explore the Missouri river, and such principal streams of it as . . . may offer the most direct and practicable water communication across the continent for the purposes of commerce. You will take careful observations of latitude and longitude, at all remarkable points on the river . . . The variations of the compass too, in different places, should be noticed.[31]

There follows a long section on knowledge to be acquired about the native inhabitants, including their languages, traditions, monuments, food, clothing, diseases and remedies, the state of morality, religion, and information among them; orders to observe

> the soil and face of the country, its growth and vegetable productions, especially those not of the United States; the animals of the country generally, and especially those not known in the United States; the remains or accounts of any which may be deemed rare of extinct [again, the hope to find a mammoth?] . . . [The] climate, as characterised by the thermometer . . . the dates at which particular plants put forth or lose their flower or leaf . . .[32] [We are back in Jefferson's garden.]

About two weeks after signing his instructions to Lewis, the news reached Jefferson that the treaty transferring the Louisiana Territories to the United States had been executed. This development, Jefferson said, "increased infinitely the interest we felt in the expedition, and lessened the apprehensions of interruptions from other powers."[33] But in any case, Captain Lewis's bags had already been packed, whether the transfer would succeed or not, and Lewis left Washington on 5 July 1803 on the first leg of his journey, to return from it only more than three years later. There is little doubt such an expedition would have been attempted, even without the treaty.

While awaiting the return of the explorer, Jefferson did what one would expect of him. He began to educate Congress. In November 1803, he submitted an "Account of Louisiana," based on whatever information he had been able to get from knowledgeable Westerners. By February 1806, Jefferson had received enough news from Lewis to prepare another message to Congress, stressing the new observations on Indians and on geography. To complement this educational purpose, he also sent to each Senator and Representative individually five reports by scholars and explorers concerning the tribes, geography, and meteorological observations of areas west of the Mississippi.

Jefferson greatly enjoyed the first shipment, in 1805, of samples from the expedition containing skins and skeletons of animals, some live animals, seeds, sixty-seven specimens of minerals and sixty plant specimens. He arranged to distribute most of these personally, some to the American Philosophical Society in Philadelphia to be studied there, to Charles Willson Peale's museum, and to expert gardeners who would grow the seeds. A few items Jefferson kept for display at Monticello.

When Lewis and Clark returned, bearing quantities of unknown flora and fauna, maps drawn by Clark, and other trophies, Jefferson constantly helped to make the results known and to speed the publication process. He treasured particularly the vocabularies of Indian languages which Lewis had collected and given to him, and it was one of Jefferson's most heartrending losses when these, together with Jefferson's notes on them, were stolen and thrown by the thieves into a river. Only a few pages were found, and Jefferson mourned over them.

In a letter to one of his scientific friends, Benjamin Rush, Jefferson had shared his joy of obtaining authority from Congress in February 1803, "to undertake the long desired object of exploring the Missouri and whatever river, heading with that, leads into the Western ocean."[34] That, and not the fur trade or the "course of empire," fashioned his project of exploration and research motivated by the joint needs of science and of the young country: a research program by which science serves both the search for truth and the interest of society.

We noted earlier that this style of research is still struggling to come to prominence. But there are indications from the history of twentieth-century science that such attempts continue. I have discussed elsewhere two examples of this Jeffersonian program (or, as it may also be called, "combined-mode" research): One was Orso Mario Corbino's description in 1929 of the need for research on nuclear physics in Italy under Fermi, both because it was the major research frontier in which success might allow Italy to "regain with honor its lost eminence" in physics, as well as making eventual power generation likely.[35] The other case was President Jimmy Carter's and his Science Adviser's initiatives in the late 1970s to identify important questions of basic scientific research which, if answered, would also have the promise of early national utility.[36]

As our century is drawing to a close, science policy confronts the clash between the necessarily increasing costs of doing important research and the ever more limiting financial constraints. When one listens to the debates in Congress, industry, and academe about the legitimacy of science to continue along the older modes, one begins to realize that—if only for the reason of renewing an endangered mandate—what we have called here the Jeffersonian style of scientific pursuit is likely to emerge as a sound contender for the attention of scientists and policymakers, as an addition which enlarges the present base of research and development.

We shall then have three types of inquiry in science and engineering. Each will have its own criteria for the evaluation of its potential, among which scientific merit and feasibility will be the most prominent. Some projects will undoubtedly be of overlapping styles, as we know from the fact that, more and more, the boundaries between basic and applied research are eroding, owing to the increasingly obvious mutual dependence between advances in basic science and in technology. And it will require sound political and scientific statesmanship to prevent enlargement of the base occurring only at the cost of reducing the already insufficient funding of the more conventional types of research. But the present course of discussions among statesmen and scientists allows one to hope for the opening at long last of the Jeffersonian route to new and useful knowledge.

Notes

1. Letter to M. Besso, 6 January 1948. Earlier, on discovering Mach's rejection of relativity, Einstein had remarked: "Mach's system studies the relationships between facts of experience; for Mach science is the totality of these relationships. That is a wrong point of view; in short, what it achieves is a catalogue, and not a system." My translation from the French in A. Einstein, *Bulletin de la Société française de Philosophie,* 22 (1922): 111.

2. Julian P. Boyd, ed., *The Papers of Thomas Jefferson,* vol. 14 (Princeton: Princeton University Press, 1958), p. 561.

3. Albert Ellery Bergh, ed., *The Writings of Thomas Jefferson,* vol. 9 (Washington, D.C.: Thos. Jefferson Memorial Association of the United States, 1907), p. 358.

4. Thomas Jefferson, letter of 28 January 1797 to the Society, *Transactions of the American Philosophical Society,* vol. 4 (1799): xii–xiii; see also Gilbert Chinard, "Jefferson and the American Philosophical Society," *Proceedings of the American Philosophical Society,* 87 (1943–1944): 263–276.

5. Bergh, ed., *The Writings of Thomas Jefferson,* vol. 13, p. 124, letter of 21 January 1812.

6. Ibid., vol. 1, p. 3.

7. See, for example, the collection of essays, *Thomas Jefferson and the Sciences,* ed. I. Bernard Cohen (New York: Arno Press, 1980).

8. Boyd, ed., *The Papers of Thomas Jefferson,* vol. 14 (1958), p. 412.

9. Ibid., vol. 16 (1961), pp. 387–389.

10. Ibid., vol. 6 (1952), p. 216.

11. Ibid., vol. 6 (1952), p. 258.

12. Ibid., vol. 8 (1953), p. 233.

13. Silvio A. Bedini, "Thomas Jefferson," in C. C. Gillispie, ed., *Dictionary of Scientific Biography* (New York: Charles Scribner's Sons, 1973), vol. 7, pp. 88–89.

14. John C. Greene, *American Science in the Age of Jefferson* (Ames, Iowa: The Iowa State University Press, 1984).

15. Thomas Jefferson, "A Memoir of the Discovery of Certain Bones of an Unknown Quadruped, of the Clawed Kind, in the Western Part of Virginia," *Transactions of the American Philosophical Society,* 4 (1799): 246–260.

16. Ibid., p. 258.

17. *Sir Isaac Newton's Mathematical Principles,* rev. Florian Cajori, original trans. Andrew Motte, 1729 (Berkeley, CA: University of California Press, 1960), p. xviii.

18. Ibid.

19. Max Planck, *Where Is Science Going?,* trans. James Murphy (New York: W. W. Norton & Co., 1932).

20. Francis Bacon, *The New Atlantis,* in *The Works of Francis Bacon,* ed. J. Spedding, R. L. Ellis, and D. D. Heath, vol. 3 (London: Longmans, 1857–59), p. 156.

21. Bergh, ed., *The Writings of Thomas Jefferson,* vol. 18, p. 160.

22. Boyd, ed., *The Papers of Thomas Jefferson,* vol. 6, p. 371.

23. Donald Jackson, ed., *The Letters of the Lewis and Clark Expedition with Related Documents, 1783–1854* (Urbana, Ill.: University of Illinois Press, 1962), pp. 669–670.

24. Greene, *American Science,* p. 196.

25. Ibid.

26. Ralph B. Guinness, "The Purpose of the Lewis and Clark Expedition," *Missouri Valley Historical Review,* 20 (1933): 90–100.

27. Bernard DeVoto, *The Course of Empire* (Boston: Houghton Mifflin Co., 1952), p. 411.

28. *The Debates and Proceedings in the Congress of the United States, Seventh Congress, Second Session* (Washington, D.C.: Gales and Seaton, 1851), p. 26.

29. Ibid.

30. Jackson, ed., *The Letters of the Lewis and Clark Expedition,* pp. 16–17.

31. Ibid., pp. 61–62.

32. Ibid., p. 63.

33. Bergh, ed., *The Writings of Thomas Jefferson,* vol. 18, p. 157.

34. Jackson, ed., *The Letters of the Lewis and Clark Expedition,* pp. 18–19.

35. G. Holton, *The Scientific Imagination: Case Studies* (New York: Cambridge University Press, 1978), pp. 164–165, and references cited therein.

36. G. Holton, *The Advancement of Science, and Its Burdens: The Jefferson Lecture and Other Essays* (New York: Cambridge University Press, 1986), pp. 188–194.

5

The Controversy over the End of Science

Even while science has been asserting ever greater success in its aim of encompassing the understanding of all natural phenomena, antithetical forces have been gathering outside the laboratory in what amounts to an effort to delegitimate science as we know it. At different times in modern history this challenge to the role of science in culture has assumed different forms; but its roots are ancient and robust.

How might one think and act in response? Useful precedents exist. This chapter focuses on the confrontation between the two main, thematically opposed positions: one claims that the sciences are by their nature subject to eventual decay; the other argues that the sciences are destined to merge eventually into one coherent body of understanding of all phenomena.

To most scientists today, the first of these choices seems too unreasonable to be taken seriously; they are unlikely to pay any attention to currently fashionable writings claiming that science, traditionally the continuing source of new insights, of material progress, and of intellectual emancipation, may now be coming to a close—to its "end"—not merely to the recognition of the limits on the power of science, limits of which scientists themselves on the whole are quite aware.[1] But to the historian of science, a debate on the possible decay and death of science is neither a contradiction nor a novelty. The idea has been proposed many times in the past. To give one example, toward the end of the nineteenth century, a number of new problems could not be solved by the then-current mechanistically based physics. In disappointment, the European scientist Emil Du Bois–Reymond wrote that science had at last come up against unbreakable barriers of understanding, beyond which we shall always remain ignorant. The cry "Ignorabimus" was raised, and soon it was converted into the more exciting slogan, "the bankruptcy of science." It spread quickly, urged on by some philosophers of science who demanded that scientists should be able to discover through their research the ultimate metaphysical reality behind phenomena. The whole epidemic collapsed when the presumably bankrupt science suddenly gave rise to such advances as quantum theory and relativity.

Since we can count on the persistent recurrence of this fascination with the idea that science could cease, our task is to learn how best to think about this topic as a whole, how to think about the possibility of an eventual end to science. And here history will help us. For with very few exceptions, virtually all proposals to this effect are driven by just one or the other of two fundamental thematic ideas.

One of these represents science as evolving essentially along a meandering, but on the whole rising, line. It recognizes the existence of occasional plateaus, even temporary downturns, but it also sees sprints of exponential growth. So on the average there is a more or less steady increase in the state of scientific knowledge, its coverage, its internal coherence, its accuracy of predictions, its refinement of the values of the natural constants. All these increases hint at the evolution in time of the kind of unified science *(Gesamtwissenschaft)* that inspired Mach's circle and its successors (see Chapter 1). The opposing view is that of scientific understanding rising for a time, but then falling and decaying, in a cyclical manner. The adherents of the first view, whom one may call the "linearists," tend to come out of the background of having actually done research in natural science. They see science as largely an autonomous activity, not primarily driven by external forces. A typical image that emerges from their writing is science as an advancing river system, branching and combining again, as it makes its way toward some total, holistic understanding of the natural world.

The "cyclicists," on the other hand, tend to think of science not as a goal-directed, progressive, cumulating activity. They are apt to base their image of the cycles of science on the biological metaphor of an organism, which develops from childhood and youth to old age and death, or the closely related political metaphor of periods of revolution, followed by a normal state, followed by yet another revolution, leading to yet another incommensurable state—a sequence of paroxysms or changes of mind that leave no certainty for certifiable progress. Cyclicists are more often to be found among social scientists and historians, and, contrary to linearists, they see science to be significantly or even mainly driven by social processes. In the extreme case, they think of science as just one expression of some general spirit of the time or even chiefly a matter of "social construction," not essentially different from the game of chess.

As is generally the case with thematically opposed positions, one cannot expect to decide for one and against the other by some simple test. Moreover, they correspond to and may result from quite opposite types of fundamental visions about the destiny of mankind itself—for the cyclicists an acquiescence in the inevitable decay of material identity, for the linearists an assertion that transcendence, a "way out" of the cycle, is possible. There may be here a resonant connection with the different conceptions of time familiar to historians of religion, e.g., the linear development implicit in Christian interpretations of historical time *versus* the cyclical ones in Eastern religions and in myths. It is not the purpose here to pursue such an analysis. Instead, we shall try to set forth these two scenarios by presenting the arguments of one of the most eloquent proponents for each side. An encounter with two interesting minds may help one to understand better what to do with this question of the putative end of science at the intellectual level.

Before being captivated by either one of these, we should take at least a moment to acknowledge the more emotionally based motivations behind the periodic calls for a stop to science. They are not the result of a rational assessment of what life would be like if continued research really were to cease. For under such conditions, mankind would not simply settle smoothly into a life of simplicity, as in some Polynesian paradise, or return to an agrarian Eden. Instead, mankind is likely to be facing almost unimaginable catastrophes, since our planet is not in equilibrium, and current knowledge is insufficient to assure a sustainable future. But the chief animating force for these flights from reason—to be discussed in detail in Chapter 6—is a fear, deep down, that the ever more rapid sequence of scientific advances, and the emancipating effects they have had from Copernicus on, have deprived much of the population of some of the instinctive bases of self-confidence, brushing them aside as superstitions. At the same time science has increased through technological means the scale and potential for damage which our violent instincts can cause us to inflict on ourselves. They are of course valid concerns, and they are being seriously addressed by scholars and scientists.

Perhaps the most direct insight into the intellectual base of the cyclicist school of thought of the fate of science can be reached through what remains to this day one of the most fascinating and outrageous of books, a book completed after ten years of labor by an obscure and impoverished German high school teacher, then in his thirties, with a doctorate in Greek mathematics and an encyclopedic ambition. These 1,200 pages, much of it written by candlelight during the First World War, offered a grand Teutonic theory of both the past and the future course of all history, interspersed with dramatic predictions, a good share of absurd-sounding speculations, and some shrewd insights. But the arresting overall conclusion of his book was revealed even by its original title, *Der Untergang des Abendlandes,* the sinking away, the annihilation of all Western civilization, including its science. The subsequent English edition gave the title of the book only inadequately as *The Decline of the West.*[2] The author's name was of course Oswald Spengler.

This enigmatic work, published in July 1918, just as the terrible war was grinding to its bitter ending, was an immediate sensation, an irresistible challenge. The debate about it, in which scientists also joined, continued for decades.[3] In his critical study of Spengler, the historian H. Stuart Hughes observed that despite all its shortcomings, and even because of them, "the book remains one of the major works of our century, the nearest thing we have to a key to our time."[4] And indeed in it we find in stark and extreme language precursors of today's arguments, familiar from the writings of Arnold Toynbee, Spengler's direct successor, and from the works of Theodore Roszak, Charles Reich, the last books of Lewis Mumford (who acknowledged his debt to Spengler), the so-called New Age authors, even some of the writers on radical feminist science, and the anti-science movements to be noted in Chapter 6.

Spengler's key conception is that for every part of mankind, in every epoch, history has taken fundamentally the same course, obeying the same morphology. And from that inevitable course follow naturally the specific forms of activity, whether social, political, literary, artistic, spiritual-religious, or indeed scientific. Each of the mighty cultures of mankind—for example, the ancient Indian, Chinese, Arabian, and the classical Greco-Roman—was not only as valid and significant as is our own Western civilization, but each is a drama with analogous structure. That is, each goes through the same season-like cycle, from its own nascent spring to its eventual burial in its own winter. Thus our own inevitable destiny in the West is to go to dust according to a timetable that can be calculated from the available precedents. Our time, Spengler said, corresponds not to that of Athens in the time of Pericles, but to that of Rome under the brutal Caesars. As it happens, we are very near the end of our cycle. Of great painting, music, architecture, or science, there can be for us no longer any hope. Our best strategy, he says, is to be bravely resigned and try at least to get a first glimpse of the rise of the next wave, which is coming from the East to triumph over the West.

Spengler tells us how each cycle progresses, from start to finish. Following Nietzsche, Spengler declares that each beginning is characterized by what he calls the Apollinian spirit, symbolized by the sensuous, individual body we can see in classical Greek sculpture. With it goes a world view embracing attention to form and the organic, rather than to the mechanical or mathematical interpretation of experience that takes its place later. A beginning is a time of contemplation, not yet of investigation, of faith rather than skepticism, of high art rather than what he calls merely the "cult of science."

At some point into this cycle, however, there occurs a kind of historic change of phase of the Apollinian soul and of the culture which it animates. It gives way to its opposite, a so-called Faustian one, which starts with a rather Germanic form of lonely romanticism, a yearning for the infinite, but gradually becomes more and more intellectualized. Thereby what was a "culture" is changed into a mere "civilization." What counts now is the notion of causality instead of destiny; attention to cause and effect rather than what

Goethe had called a "living nature"; to abstractions such as infinite and empty space rather than the palpable earth which you can feel and smell. In a civilization, the primacy of soul is replaced by intellect; concern for human needs degenerates into debates about money; mathematics pervades more and more activities; the principle of causality is forced on the understanding of phenomena; and nature is reinterpreted as a network of laws within the corpus of "scientific irreligion."

This transition from culture to civilization was completed in the fourth century for the world of antiquity in Europe; Spengler proposes that the same transition began in the late nineteenth century for the cycle of our Western society. As in past cycles, the phase in which we find ourselves will not end abruptly. It will linger on for some time. We have entered the last stage also in world politics, marked by the replacement of "the idea of the state's service" by the naked "will to power." As Nietzsche had predicted, ours would be the century of tyrants, of weapon-hungry Caesars engaged in a struggle for world rule—even as offstage an entirely new culture is getting ready to take over the field.

It is of special interest to us that in Spengler's somber drama, science plays a crucial role. The Faustian element in science, Spengler informs us, is exemplified succinctly by the famous confession of the physicist Hermann von Helmholtz, who wrote that "the final aim of natural science is to discover the motions underlying all alterations, and the motive forces thereof; that is, to dissolve all natural science into mechanics." This urge is not merely an expression of the universal longing to find the One in the Many. More specifically, Spengler notes, in our science "the *seen* picture of nature [is converted] into the *imagined* picture of a single, numerically and structurally measurable order." If he were writing today, Spengler would perhaps have replaced Helmholtz's quotation with the recent one by the physicist Leon Lederman, who, encouraged by the current success of the unification program in physics, has mused that the aim of science now is to reduce the laws governing all natural phenomena to one equation that will fit on a T-shirt.

Now Spengler introduces his most startling idea, one that has become familiar in new garb also. He warns that it is characteristic of the winter phase of civilization that precisely when high science is most fruitful within its own sphere, the seeds of its own undoing begin to sprout. This is for two reasons: the authority of science fails both within and beyond its disciplinary limits, and an antithetical, self-destructive element arises inside the body of science itself that eventually will devour it.

The failure of science's authority outside its laboratories, he says, is due in good part to the tendency to overreach and misapply to the cosmos of history the thinking techniques that are appropriate only to the cosmos of nature. Spengler holds that the thought style of scientific analysis, namely "reason and cognition," fail in areas where one really needs the "habits of intuitive perception" of the sort he identifies with the Apollinian soul and the philosophy of Goethe.

But, Spengler adds, even in the cosmos of nature there is an attack on the authority of science, arising from within its own empire. For every conception, even in science, is at bottom "anthropomorphic," and each culture incorporates this burden in the key conceptions and tests of its own science, which thereby become culturally conditioned illusions.

For example, "to the Classical [cycle] belonged the conception of form; to the Arabian, the idea of substances with visible or secret attributes; to the Faustian—ours—the ideas of force and mass." In particular, the Faustian physics of the last 300 years has been a physics of dynamics and of "methodical experiments," both of which, Spengler says, are exemplifications of the will to power that imbues the civilization phase of a people, when "Nature is not merely asked or persuaded, but forced." All our rushing after positive scientific achievements in our century only hides the fact, he thinks, that as in Classical times, science is once more destined to "fall on its own sword" and so make way for the coming world outlook, what he calls the "second religiousness." Indeed, guided by his theory of cycles, Spengler tells us "it is possible to foresee the date when Western scientific thought shall have reached the limits of its evolution." And in one of the handy chronological charts which Spengler put at the end of his book, he allows us at the same time to see his millenarian roots and to find that fateful date. It is the year 2000.

Indeed, to Spengler's eyes the signs of decay and disintegration in science were clear already by 1918. Physics, he says—and note how familiar this too has become in recent Dionysian works about science—physics has been infected by an "annihilating doubt," as shown by "the rapidly increasing use of statistical methods, which aim only at the probability of results and forego in advance the absolute scientific exactitude that was a creed to the hopeful earlier generations." The possibility of a self-contained, self-consistent mechanics has to be given up because "the living person of the knower methodically intrudes into the inorganic form world of the known." Moreover, the ruthlessly cynical hypothesis, as he calls it, of the relativity theory strikes at the very heart of dynamics. The quantum ideas are held to be equally destructive. And Spengler adds that he is alarmed at "how rapidly card houses of hypotheses are run up nowadays, every contradiction being immediately covered up by a new hurried hypothesis." So, turning away from the search for exactitude and absolutes and adopting probabilism have undermined science from the inside. Our inability, for example, to specify which atom in a sample of radioactive material will decay next points directly to the Achilles' heel of modern science. It is as if the idea of destiny instead of causality has been unwittingly reintroduced into the picture of nature.

And yet another, final cause for the self-destruction of the modern scientific world picture arises, he says, from its tendency to theory and to symbol orientation. For what is happening is that all the separate sciences are converging into one, a "fusion" characterized by the reduction to "a few grand formulas" in the winter of science. But ironically, just this has led us now back precisely to what is the first and simplest activity in the beginning of every new culture, what is always part of its primitive religious spirit: that is, the preoccupation with numerical regularities. Number is part of the earliest religious belief and ritual; number mysticism appears in every faith in such sacred concepts as the relation of microcosm to macrocosm or in the building of prehistoric structures that served both for religious rites and for astronomy.

All these internal cancers will shortly kill science as we know it, and we shall rediscover that at bottom mankind as a whole, he says, has never wanted to analyze and prove, but has only wanted to believe. What he calls this orgy of three centuries of exact sciences is ending, together with the rest of what was valuable in Western civilization. Indeed, the only activities which are on the ascent during this final act are economics, politics, and technology. And as a kind of postscript, in his later book, *Man and Technics* (1931), Spengler adds his opinion that advancing technology, with its mindlessly proliferating products, will also turn out to undermine the society of the West because, according to his uncanny prediction, there will be a failure of science and engineering education: the level of teaching in the metaphysically exhausted West will not be up to maintaining technological advance. The attraction of the scientific-technological profession is diminishing. "The Faustian thought begins to be satiated with machines . . . and it is precisely the strong and creative talents that are turning away from practical problems and sciences . . . Every big entrepreneur has occasion to observe a falling-off in the intellectual qualities of his recruits." At the same time, the previously overexploited races, "having caught up with their instructors," have begun to surpass them, to "forge a weapon against the heart of the Faustian Civilization." The non-Caucasian nations will adopt the technical arts and turn them against the Caucasian inventors. One of Spengler's commentators simply summarized Spengler's prediction of 1931: "Already they can undersell the products of Western industry. Eventually they will conquer the Western nations themselves."[5]

Thus spoke the ancestor of the end-of-science movements. It is obviously rather easy to find specific faults with this work, as it is with the derivative versions that clamor for attention today. Prominent among the deficiencies one cannot fail to note the presence of distorted versions of Hegelian and Marxian dialectics, and more specifically the frequent, basic misunderstanding about science by Spengler and his heirs. For example, the use of probability and of quantum causality is not an abandonment of all causality as such. The notion of entropy does not, as he thought, inevitably lead to the heat death of the universe. The subjectivity of the individual does not rob science of all claims to objectivity. And so on. Moreover, Spengler, who was really a nineteenth-century thinker, could not have foreseen the rapid internationalization of almost every aspect of science. Even if the Occident should in some deep sense eventually decay and some other culture takes its place, it is a safe bet that, short of a return to total primitivism, the new schools will also be teaching Euclid's geometry, Harvey's blood circulation, Newtonian dynamics, Einstein's space-time, Norbert Wiener's cybernetics, and the Watson-Crick double helix. These wheels cannot be un-invented.

On the other hand, one must credit Spengler with the perceptive insistence that despite what he called the "irreligiousness" of science, there is a subterranean link between science and religion at their origins. And that particular, unpopular aspect of Spengler's cyclicist view had some analogy in the work of a very different person. It is in fact the person we have chosen to represent now the opposite, the linearist view of the fate of science. For this purpose, one could well have turned to the writings of other scientists, such as Johannes Kepler or Hans Christian Oersted or Niels Bohr. But it is more appropriate to select as the linearist exemplar an essay that also appeared in 1918, within a few months of Spengler's book, written by a man of almost the same age as Spengler, and one who was then also still almost unknown outside his own circle. That essay was originally a speech given in honor of the sixtieth birthday of Max Planck, whose work Spengler had just found to be destructive to science. And the name of the young speaker, whose work Spengler had also singled out as a symbol of disintegration, was Albert Einstein.

Rising at that dark point in European history, Einstein began his short but memorable talk[6] with an image: "The temple of science is a vast building with many different wings." In it, many are there who pursue science out of the joy of flexing their intellectual muscles, and others for short-term utilitarian ends. But, happily, there are also a few who do it simply because of their deep longing for knowledge itself. What led those few into the temple? They have two motives for doing science. One is negative—a desire to escape one's "everyday life with its painful harshness and wretched dreariness, and from the fetters of one's own shifting desires."

But there is also a positive motive. "Man seeks to form, in whatever manner is suitable, a simplified and lucid image of the world, a world picture," a coherent view of how the cosmos of experience hangs together, "and so to overcome the world of experience [a Schopenhauerian concept], by striving to replace it to some extent by that image. That is what painters do and poets and philosophers and natural scientists, all in their own way. And into this image and its formation each individual places his or her center of gravity of the emotional life, in order to attain the peace and serenity which cannot be found within the confines of swirling personal experience."

The picture of the world which the physicist is building is only one among all the other possible ones. But "it demands rigorous precision in the description of relationships." Therefore, the physicist must be content with studying first an idealized world, where for example all friction is negligible. "This allows him to portray the simplest occurrences which can be made accessible to our experience." The more complex phenomena of the real world cannot be immediately attacked with the necessary degree of logical perfection and accuracy. Therefore, at the beginning of a problem the scientist strives for "supreme purity and clarity, but at the cost of completeness."

This simplifying reductionism—to which the Romantic critics, from Goethe through Spengler to this day, are so opposed—is only the first, preliminary stage in Einstein's theory of scientific advance. History has taught us, he continues, that once a world image has been achieved on the basis of simplification, it turns out to be at least in principle extensible to every natural phenomenon as it actually occurs, in all its complexity and its completeness. Reductionism is only a detour to the road leading to the eternal, synthetic laws.

And now, going beyond the *Bildungs* ideal of German intellectuals of the period, Einstein reveals the long-range agenda for science as he sees it, the destiny of science: from the general laws "it should be possible to obtain by pure deduction the description, that is to say the theory, of *every* natural process, including those of life." That promise of the eventual unification of all exact knowledge is the final aim, the *telos* toward which Einstein sees science striving.

We may well note here that in fact in the intervening years enormous progress has been made in this direction—for example by finding that a good deal of chemistry is just that part of atomic and molecular physics which really works; by discovering the bridge between biological and physical sciences via DNA; by finding some deep links between aspects of behavior and one's genetic endowment or biochemical imbalance—and in the program of the unification of the forces of physics. In short, in modern form the old theme of finding the One in the Many has become the stuff of which Noble Prizes are made. It is no longer entirely the dream of Faust, who in Goethe's drama exclaimed that either he would attain the knowledge of everything, or else he would have to remain a mere worm.

But to return to Einstein's talk. At this point he issues the warning that the general project for the eventual unification of all the sciences, while yielding ever deeper insights and being a powerful motivation, is likely to be one without an early or foreseeable end. The meandering line tracing out the advance of science is not terminating; we may have an infinite task on our hands. One reason is that despite all our successes we really lack a reliable method or guaranteed algorithm, for we have to make do with the fallible capacities of human thinking. Far from embracing the stereotype of a relentless victory march of cold rationality, which in any case exists only in bad science textbooks, Einstein freely confesses here, as he was to do again and again later, and contrary to the then-reigning philosophy, that "to the [grand] elemental laws there leads no logical path, but only intuition."

Of course that does not mean that anything goes, or that science has lost its authority and is doomed to stumble blindly from one discovery or system of theories to the next. While there is no logical bridge from experience to the basic principles of theory, and hence no proof of the validity of philosophical realism itself, in practice we have good tests for the degree of veracity of our theories. In addition, there is the fact, the astonishing fact, that agreement is possible within the very heterogeneous scientific community. That is a sign that "the world of experience does uniquely define the theoretical system." Even though *a priori* we had no right to expect any such correspondence, somehow the order we put into our theories can, and often remarkably does, turn out to correspond to the order others find in nature when they check our predictions.

Why is that possible? Why can our limited mind penetrate so often and so well behind the appearances to discern a few universally valid laws? How can it find its way back and forth between the world of phenomena and the world of ideas? On that point, Einstein confesses freely, he has no certain answer. But that does not make him collapse in demoralized helplessness. He has a daring suggestion—that our minds are guided by "what Leibniz termed happily the 'pre-established harmony.'"

Gottfried Wilhelm Leibniz, the philosopher and contemporary of Newton, had postulated that our ability to discover the laws concerning material bodies is one aspect of the unity from which God created the two apparently separate entities of the universe, the spiritual and the material. Each of these obeys its own laws; but they can interact in sympathetic unison, somewhat in the way a stringed instrument goes into resonance and picks up the sounds made by a second one that is tuned to it. Or, to use Leibniz's own words to explain this possibility of a harmonious interaction, words that must have delighted Einstein: "The souls follow their laws . . . and the bodies follow theirs . . . Nevertheless, these two beings of entirely different kind meet together and correspond to each other like two clocks perfectly regulated to the same time. It is this that I call the theory of pre-established harmony."

Scientists of our day are more likely to invoke an argument from the supposed evolutionary base of a correspondence between our ideas and our environment. They will do so less because of any proof and more because they feel uncomfortable with the theological undertone of Einstein's metaphor, one which would have come more naturally to those who, like him, were familiar with Leibniz's discussion from their reading of the commentary on it in the writings of Immanuel Kant. But to Einstein just this undertone was by no means unwelcome or accidental. Having nearly reached the end of his essay with this image, Einstein returns briefly to the question of what motivates people to pursue science despite the lack of any guarantee of success or even of an end to their labors. It is wrong, he concludes, to trace this persistence "to extraordinary willpower or discipline." Rather, "the state of feeling which makes one capable of such achievements is akin to that of the religious worshipper, or of one who is in love. [That is,] one's daily strivings arise from no deliberate decision or program, but out of immediate necessity."

In the years that followed, Einstein continued on every occasion he could find to spell out and develop these views: Science is a program with an aim toward which one can advance, but it has no ending in the foreseeable future. It is a mandate to produce the best objective description possible of the physical cosmos, while having to work only with one's subjective capacities and with essentially arbitrary concepts. It is an activity of persons able to combine logical rationality with intuition (contrary to the Spenglerian assumption of their incompatibility), who have the knack for advancing both on hard evidence and on faith, and sometimes even on aesthetic grounds. Doing science requires analysis as well as synthesis. In short, science is the mobilization of the whole spectrum of our talents and longings, in the service of shaping more and more adequate world pictures. What to lesser minds looks like a mixture of mutually exclusive opposites between which one must make a choice, to Einstein seemed to be complementary necessities.

It is therefore not surprising that he, unlike Spengler and his followers, also saw no inherent conflict between science and religion, as Einstein hinted in this passing reference to the kinship between the scientist and the religious worshipper. In later essays[7] he elaborated his deeply felt argument that scientific activity, the search for the evidence of rationality in the universe, is in essence a "religious act." As one would expect, his description of what he called "Cosmic Religion" is not a product of sentimentality or of sectarianism; nor do religion and science, where they merge into Cosmic Religion, have much in common with the conceptions held dear by any religious establishment. Einstein's idea of God was not that of the biblical, intervening deity. Rather, his view, derived in part from Spinoza, serves as a necessary reminder that science, from its earliest beginnings to our time, has retained the signature of that single, undifferentiated totality which motivates our inherently endless human search both for explanation and for transcendence.

A few analogies can be noticed between Einstein's linearist views and those of the cyclicists such as Spengler and his followers. For example, Einstein too was opposed to the more imperialistic claims of positivism. But the essential, overriding difference between them is that for Einstein, as for most modern scientists, the notion of a foreseeable ending of science is a contradiction in terms, and there is no evidence to the contrary. For them, doing science "out of immediate necessity," with neither a determinate timetable nor guaranteed algorithms, is inherently a rather turbulent activity, one well captured in a memorable analogy in Otto Neurath's essay "Antispengler":[8] "We are like sailors who on the open sea must reconstruct their ship but are never able to start afresh from the bottom . . . They make use of some drifting timber of the old structure, to modify the skeleton and the hull of their vessel. But they cannot put the ship in dock in order to start from scratch. During their work they stay on the old structure and deal with heavy gales and thundering waves . . . That is our fate."

This picture of science as an incessant, self-constructing enterprise against great odds was improved by the philosopher Hilary Putnam:[9]

> My image is not of a single boat but of a "fleet" of boats. The people in each boat are trying to reconstruct their own boat without modifying it so much at any one time that the boat sinks . . . In addition, people are passing supplies and tools from one boat to another and shouting advice and encouragement (or discouragement) to each other. Finally, people sometimes decide they do not like the boat they are in and move to a different boat altogether. And sometimes a boat sinks or is abandoned. It is all a bit chaotic; but since it is a fleet, no one is ever totally out of signalling distance from all the other boats. We are not trapped in individual solipsistic hells (or need not be), but invited to engage in a truly human dialogue, one which combines collectivity with individual responsibility.

As we look back on this confrontation between two contemporaries representing in extreme form the two most widely held, opposing theories about the eventual fate of science, it should be clear that they do not encompass all positions possible from our *fin-de-siècle* standpoint. To mention just one divergence, a small but growing group of scientists appears now to be quite comfortable with a style of work that is on neither the linearist nor the cyclicist trajectory but opts frankly for an inherent pluralism. They disclaim any expectation for an ultimate coherence of all parts even within a given science. These might be called splitters rather than lumpers. They have an important role in the advancement of science, for that often depends on the interaction and alternation of these two traits of research—as if science moved on two feet. This point was put well, in the context of his time, by the Danish scientist Hans Christian Oersted:[10]

> One class of natural philosophers has always a tendency to combine the phenomena and to discover their analogies; another class, on the contrary, employs all its efforts in showing the disparities of things. Both tendencies are necessary for the perfection of science, the one for its progress, the other for its correctness. The philosophers of the first of these classes are guided by the sense of unity throughout nature; the philosophers of the second have their minds more directed towards the certainty of our knowledge. The one are absorbed in search of principles, and neglect often the peculiarities, and not seldom the strictness of demonstrations; the other consider the science only as the investigation of facts, but in their laudable zeal they often lose sight of the harmony of the whole, which is the character of truth. Those who look for the stamp of divinity on every thing around them, consider the opposite pursuits as ignoble and even as irreligious; while those who are engaged in the search after truth, look upon the other as unphilosophical enthusiasts, and perhaps as phantastical contemners of truth . . . This conflict of opinions keeps science alive, and promotes it by an oscillatory progress.

A second "minority" type of divergence from the two main models for the fate of science is represented by the belief of the physicist P. W. Anderson.[11] Anderson sees a "hierarchical structure of science" that does not permit in principle a reduction to one set of fundamental laws from which one could then "reconstruct the universe." For example, the problems of scale and of complexity do not allow the properties of large aggregates of elementary particles to be understood merely by extrapolation of the behavior of individual particles. Rather, by a process analogous to the old conception of "emergence," in each level of complexity there can be imagined to arise entirely new properties; hence each is likely to have a conceptual structure of its own, and presumably also its own rate and direction of progress. These are quite contrary to the ideals of the linearists, for whom the hierarchical structure of science does not separate the layers of science but rather helps to fix the direction of the arrow of fundamentality that points to the discovery of ultimate laws of nature. Thus Steven Weinberg wrote that "Nature has absolute laws of great simplicity, from which all the sciences flow in a hierarchy."[12]

As suggested earlier, there is little hope of deciding in the abstract which of the various models of scientific progress will prevail in the long run. Yet on present evidence one can predict that most active scientists will continue to take greatest exception to the cyclicist model, with its notion that science has exhausted its mandate. They will at best be bemused to hear that scientific progress is now thought in some quarters to be intellectually indefensible, an idea "in crisis." Ignoring such claims, they will continue to hold that it is the particular mission and talent of scientists, as of others, to seek certifiable truths with whatever limited means come to hand; that apologies are required neither for their impulses to seek rational meaning in those signals that reach them, nor for the innate tendency to seek transcendence even in science; and that their untidy mixture of motives, their unguaranteed tools, and their open-ended program will continue to grip them as they rebuild their ships against the perils of the ocean.

Notes

1. For example, the Nobel Conference XXV, held in October 1989, contained the following agenda-setting paragraphs in its letter of invitation to the participants of the conference:

"As we study our world today, there is an uneasy feeling that we have come to the end of science, that science, as a unified, universal, objective endeavor, is over. Even the consensus that science is a recently formed alliance, a consensus that has led to the grand methodologies of science, is in fragments.

"We have begun to think of science as a more subjective and relativistic project, operating out of and under the influence of social ideologies and attitudes—Marxism and feminism, for example. This leads to grave epistemological concerns. If science does not speak about extra-historical, external, universal laws, but is instead social, temporal and local, then there is no way of speaking of something real behind science that science merely reflects."

In the same spirit, a conference was held in December 1991 at the Massachusetts Institute of Technology under the title "Progress: An Idea and Belief in Crisis"; the letter of invitation remarked that "the idea of Progress" is "predicated on the belief in reason and material advancement. The value and validity of both of these have now been seriously called into question. It is this situation that has produced a crisis in belief."

2. I shall be basing my analysis on, and quote from, the following: Oswald Spengler, *Der Untergang des Abendlandes: Umrisse einer Morphologie der Weltgeschichte,* vol. I, *Gestalt und Wirklichkeit* (Vienna, Leipzig: Wilhelm Braunmüller, 1918); Spengler, *Der Untergang des Abendlandes: Umrisse einer Morphologie der Weltgeschichte* (Munich: C. H. Beck, 1980), which contains, in revised edition, both vol. I, *Gestalt und Wirklichkeit,* and vol. II, *Welthistorische Perspectiven* (originally published 1922); Spengler, *The Decline of the West,* vol. I (New York: A. A. Knopf, 1926), and vol. II (1928); Spengler, *Der Mensch und die Technik: Beitrag zu einer Philosophie des Lebens* (Munich: C. H. Beck, 1931), translated as *Man and Technics: A Contribution to a Philosophy of Life* (New York: A. A. Knopf, 1932); Spengler, *Briefe, 1913–1936* (Munich: C. H. Beck, 1963), translated as *Letters of Oswald Spengler, 1913–1936* (New York: A. A. Knopf, 1966).

3. Some of the debate is summarized in M. Schroeter, *Der Streit um Spengler: Kritik seiner Kritiker* (Munich: C. H. Beck, 1922), and in Schroeter, *Metaphysik des Unterganges* (Munich: Leipzig Verlag, 1949).

4. H. Stuart Hughes, *Oswald Spengler: A Critical Estimate* (New York: Charles Scribner's Sons, 1952), pp. 164–165.

5. Hughes, *Oswald Spengler,* p. 121.

6. Albert Einstein, "Principles of Research" (a mistranslation of what should be "Motivations of Research"), in *Ideas and Opinions* (New York: Crown Publishers, Inc., 1954), p. 224.

7. Including three essays that also appeared in Einstein's *Ideas and Opinions.*

8. Chap. 6 in Otto Neurath, *Empiricism and Sociology* (Dordrecht, Boston: D. Reidel Publishing Co., 1973).

9. Hilary Putnam, "Philosophers and Human Understanding," in A. F. Heath, ed., *Scientific Explanation* (Oxford: Clarendon Press, 1981), p. 118.

10. Hans Christian Oersted, "Thermo-electricity," *The Edinburgh Encyclopaedia, 1830.* Reprinted in Kirstine Meyer, ed., H. C. Oersted, *Naturvidenskabelige Skrifter,* vol. 2 (Copenhagen, 1920), p. 352.

11. Cf. P. W. Anderson, "More Is Different," *Science,* 177 (1972): 393–396.

12. S. Weinberg, "Why Build Accelerators?," in Luke C. L. Yuan, ed., *Nature of Matter: Purposes of High Energy Physics* (New York: Brookhaven National Laboratories, 1965), pp. 171–172. Weinberg adds this interesting remark as a footnote: "I do not necessarily wish to imply that we can expect to find a set of ultimate physical truths within the next few centuries (though I happen to believe we shall). It may be that we shall discover an infinite regress of more and more fundamental sciences, or even that we shall pass outside the bound of science itself to some new mode of thought which we can now no more imagine than Plato could conceive of the modern scientific method. In any case, not only scientists will be interested to see what happens."

6

The Anti-Science Phenomenon

Opposition to science as conventionally defined can take a great variety of forms, from interest in astrology to attacks on relativity theory, from false beliefs based on scientific illiteracy to support of Lysenkoism or Creationism. Which of these attacks are relatively negligible, and which are dangerous? What do these symptoms of disaffection with the Enlightenment-based tradition portend for science and culture in our time? Once we have a framework to deal with the belief in anti-science (or "alternative science," "parascience"), we shall recognize that such belief is grounded in a person's functional world view; it is one symptom of a long-standing struggle over the legitimacy of the authority of conventional science, as well as of the concept of modernity within which science claims to be embedded. An analysis of anti-scientific beliefs might lead finally to the identification of a set of strategies for dealing with the countervisions that periodically attempt to raise themselves from the level of apparent harmlessness to that of politically ambitious success.

To be sure, conventional cultural analysts may give topics other than "anti-science" a higher priority in any study of the social and political dimensions of science and technology. Some academics may be drawn more to considering whether there is a link here with the spread of analogous antipathies to the Western tradition in literature and the arts. Others may see the more urgent problem facing our civilization to be the excesses of revengeful nationalism, fundamentalism, and ethnic strife, or of the celebration of violence (what Freud, in "Why War?," called the human race's *"Destruktionstrieb,"* its "destructive instinct"). In comparison with these, anti-science may seem only an ephemeral phenomenon. But in my view the topic merits serious attention, not least because it is, historically and potentially, connected ominously to those other, more obvious dangers.

The Surface of the Problem

Recently, a conference was called to help scholars in the republics of the former Soviet Union to understand and deal with the Glasnost-released flowering in their lands of publications promoting "other ways of knowing," mystics, clairvoyants, astrologers, extra-terrestrial visitors, faith healers, and the rest of the—to us—familiar cast of characters. Just as there has been a downturn of interest in pursuing science and engineering as a profession in the West, a similar attitude has become prominent in those countries. So it appears that an alarm bell is sounding on both continents, one that calls us to contemplate "how superstition won and science lost," to use the title of John C. Burnham's useful book.[1] We seem to be urged to share any knowledge that can be expected to help cure the body politic of its disease and return it to the healthy state to which we, as children of the Enlightenment, think our fellow-citizens have the right and duty to aspire at the end of this blood-drenched century: a state that is rational, progressive, anti-superstitious, pro-science, and free of the medieval curses of folk magic, miracle, mystery, false authority, and mindless iconoclasm.

However, conscience demands that I declare at the outset that I shall not try to provide a map to this paradise. First the category *anti-* will have to be reformulated if we are to grasp the problem correctly. Indeed, I see my main task to outline how to think about anti-science at the proper level. The term *anti-science* can lump together too many, quite different things that have in common only that they tend to annoy or threaten those who regard themselves as more enlightened. We must disaggregate from the disparate jumble that which is the truly worrisome part of anti-science, so that we can discriminate between "real" science (good, bad, and indifferent; old, new or just emerging); pathological science (as in Irving Langmuir's essay on people who thought they were doing real science but were misled);[2] pseudo-science (astrology and the "science" of the paranormal); blatant silliness and superstition ("pyramid power"); scientism (the overenthusiastic importation of "scientific" models into nonscientific fields; or the vastly exaggerated claims of technocrats for scientific and technological powers, such as the "Star Wars" projects); and other forms.

Thereby we shall be able to focus on the single most malignant part of the phenomenon: the type of pseudo-scientific nonsense that manages to pass itself off as an "alternative science," *and does so in the service of political ambition.* Here our Russian colleagues may be able to instruct us because of their unhappy experience in past decades with Lysenkoism, attacks on the relativity theory and quantum mechanics, and on cosmologists who were thought to have offended against the doctrines of Engels' *Anti-Dühring.* That is the general area which will call for careful attention. We must not become preoccupied with surface phenomena. For example, much of tabloid sensationalism involving UFOs is merely hucksterism feeding on primitive ignorance (unless, as with the reputed recent inauguration of a section on "UFO-logy" in the Russian Academy of Science, the craze gets official backing).

Yet, if our aim is to filter out, name, and analyze the really dangerous segment of what some call the "anti-science movement," we shall not find much help in the literature. There exists no adequate, serious treatment of it, nor even of the modern outlook that feels threatened by anti-science. All of us enter this study equally in need of a better understanding. Nor do we really comprehend the causes of one of the preconditions of false ideas, namely the rampant scientific illiteracy in the United States. There is an extensive literature on this topic; here we need only refer to a report conveyed to the Congress by the President's Science Adviser.[3] Public scientific literacy in the United States is now at a level where "half the adults questioned did not know that it took one year for the Earth to orbit the Sun" (p. 8). (As we know from other surveys,[4] less than 7 percent of U.S. adults can be called scientifically literate by the most generous definition, only 13 percent have at least a minimum level of understanding of the process of science, and 40 percent disagree with the statement "astrology is not at all scientific.") In particular, "Teaching is a profession in crisis . . . We are currently losing thirteen mathematics and science teachers for each one entering the profession" (p. 5). Only the following percentages of teachers meet the minimum established standards for course work preparation at the high school level: 29 percent in biology, 31 percent in chemistry, 12 percent in physics (p. 6). Typically, in nearly 30 percent of U.S. high schools, physics courses are not even offered (p. 5), and only 20 percent of the high school graduates have taken physics courses of any kind. "In the most recent international science assessments, in comparison with students in 12 other countries, our high school students finished 9th in physics, 11th in chemistry, and last in biology . . . In mathematics, our top 13 percent generally fell into the bottom 25 percent in comparison with other countries" (p. 25).[5]

Why Does the Anti-Science Phenomenon Concern Us?

That such a small fraction of U.S. adults can be called scientifically literate at a time when the accomplishments of modern science, the feats of technology, and the effects of both on our lives are more spectacular than ever is not merely ironic but profoundly in need of explanation. To this intellectually important point is joined a political one: in a democracy, no matter how poorly informed the citizens are, they do properly demand a place at the table where decisions are made, even when those decisions have a large scientific / technical component. In that lies the potential for erroneous policy and eventual social instability. For as I shall illustrate, history has shown repeatedly that a disaffection with science and its view of the world can turn into a rage that links up with far more sinister movements.

It is thoughts of this kind which the phenomenon of anti-science raises in the minds of many intellectuals, West and East. By themselves, all the astrologers, anti-evolutionists, spiritualists, psychics, and peddlers of New Age thinking could otherwise be merely a target of our condescension or a source of amusement. We seem to discern behind these multi-faceted phenomena—and the related illiteracy in history, geography, etc., which we shall cover for now with the delicacy of embarrassed silence—something perilous, a potentially fatal flaw in the self-conception of the people today. As we saw in Chapter 5, soon after the start of this century, Oswald Spengler taught a fascinated public that the ideas of modern science themselves contained the poison leading to the inevitable Decline of the West, by what he called "metaphysical exhaustion"; and Max Weber announced that the method of natural science was a systematic "process of disenchantment" of the world, with its resulting loss of "any meaning that goes beyond the purely practical and technical . . . [a] question raised in the most principled form in the works of Leo Tolstoy."[6] Could it be that, on having reached the end of the twentieth century, we will find that the widespread lack of a proper understanding of science itself might be either a source, or a tell-tale sign, of our culture's decline?

It would be a vast oversimplification to think this alone is an explanation of a complex social development; but one must not dismiss it as one component for our consideration. And it is not an unfamiliar position. One of the most eloquent analyses of the exhaustion or abandonment thesis, and its parallel in early history, is to be found in the last chapter, entitled "The Fear of Freedom," in E. R. Dodds's book, *The Greeks and the Irrational.*[7] The rise of the Greek Enlightenment in the sixth century B.C., following the Homeric Age, was characterized by a "progressive replacement of the mythological by rational thinking among the Greeks." But by the end of the reign of Pericles, the tide had turned again, and teaching astronomy or expressing doubts about the supernatural became dangerous. Cults, astrology, magical healing, and other familiar practices were symptoms of the onset of a long decline, which Dodds terms the "Return of the Irrational." And, Dodds asks, have we similarly entered now on the end phase of that second great experiment with rationalism, generally identified with the Scientific Revolution and the Era of Enlightenment? Is there not even a parallel here to one of the reasons for the opening of the abyss in antiquity—that "as the intellectuals withdrew further into a world of their own [from the late period of Plato on], the popular mind was left increasingly defenseless . . . and, left without guidance, a growing number relapsed with a sigh of relief into the pleasures and comforts of the primitive"?

By the late fifth century, "the growing rationalism of the intellectuals was matched by regressive symptoms in popular belief," as the gap widened "into something approaching a complete divorce." Intellectually abandoned during a sort of Decline of the

Mandarins, the masses were prey to the spread of astrology and the like, in good part because of the "political conditions: in the troubled half-century that preceded the Roman conquest of Greece it was particularly important to know what was going to happen . . . For a century or more the individual had been face to face with his own intellectual freedom, and now he turned tail and bolted from the horrid prospect—better the rigid determinism of the astrological Fate than that terrifying burden of daily responsibility," that freedom which did not lead to certainty and safety.

Who does not hear in this the thundering voice of the Grand Inquisitor of Dostoevsky's *Brothers Karamazov?*

> No science will give the masses bread so long as they remain free. In the end they will lay their freedom at our feet and say to us: "Make us slaves, but feed us" . . . There are three powers, three powers alone, able to conquer and to hold captive forever the conscience of these impotent rebels for their happiness—those forces are miracle, mystery, and authority.

One may try to shrug off such dark thoughts by pointing to the bright side, not least the practically universal popular enchantment with high-tech. One may seek comfort in the fact that even though only less than half of the U.S. adult population believes in the evolutionary descent of human beings from earlier species, and even though half has trouble finding one side of a square when given one of the other sides, the U.S. public at large reports to pollsters a greater level of belief in the potential of science and technology as a force for the good (at least in the abstract) than equivalent tests have shown for other major industrial countries, such as France and Japan.

This uninformed assertion of interest is not troubled by the well-documented, contradictory feeling about scientists, which is far less positive. In America today it is not science but religion which, as in the days of the seventeenth-century Pilgrims, is perhaps the strongest force in private and national life—just as Tocqueville had noticed in the 1830s. About one-third of our adults, and a large fraction of these from evangelical sects, now say they are "born-again" believers; over half believe in the possibility of the daily occurrence of miracles through prayer; 60 percent say they believe in the literal existence of Hell for the eternally damned. And the financial support given annually as private donations to religious organizations now amounts to well over $75 billion. But here again, there is little consciousness of any contradiction, despite the fact that the modern, science-based world view evolved in good part from a reaction to just such a contradiction, and indeed suffers still from the inability to find a way to bridge the chasm between these two undeniable imperatives, science and faith. By contrast, the large majority of average Americans reports experiencing no conflict at all between these different forces.[8]

Similarly, while ideas that one commonly takes to be anti-scientific are widespread in the U.S., there is important evidence that this, too, is not a simple or monolithic attitude. Rather, there is a coexistence of potentially opposing kinds of consciousness. And that, as we shall see, lends itself to strategies for change. Like the different tectonic plates in the Earth's crust that tend to move in opposite directions, with occasionally disastrous results, the various elements making up the mind-set of the average person today do not form a harmonious whole. As Dostoevsky's Grand Inquisitor knew, the liberal, Enlightenment-based view deludes itself if it assumes it has been victorious. Indeed, the "pro-science"-imbued world picture of the late twentieth century is a rather vulnerable and fragile minority position, the more so as scientists and other intellectuals as a group have not managed to create sufficient effective institutional or other intellectual forums for even discussing among themselves, and with others, what the powers and limits of science in these respects are. (The uneasy toehold of Science–Technology–Society studies in most major universities is just one evidence of this lack of attention.)

Anti-Science as Countervision: Forces of Delegitimation

The evidence of internal contradictions is a signal that we must submit the anti-science phenomenon to another level of analysis. To understand in more satisfactory terms what in fact is meant by anti-science, and what it may imply for the future of our culture, we must start with the recognition that no culture can be truly anti-scientific, in the sense of opposing the activity "science" (science as defined, for example, in the *American Heritage Dictionary of the English Language:* "The observation, identification, description, experimental investigation, and theoretical explanation of natural phenomena"). Although some philosophers of science will have trouble with aspects of such a definition, I do not find even the most Dionysian "anti-scientists" calling for opposition to that activity as such.

Moreover, the anti-science phenomenon is not at all just an incomplete or ignorant or damaged form of the "proper" world view that many believe should characterize our civilization at this time in history. Instead—and leaving aside the banal, relatively harmless, or ignorant varieties—what the more sophisticated so-called anti-scientists offer is, to put it bluntly, an articulated and functional, and potentially powerful, countervision of the world, within which there exists an allegiance to a "science" very different from conventional science. And that countervision has as its historic function nothing less than the delegitimation of (conventional) science in its widest sense: a delegitimation which extends to science's ontological and epistemological claims, and above all to its classic, inherently expansionist ambition to define the meaning and direction of human progress. In short, we are watching here an ancient, persistent, obstinate, and hardly ameliorating combat.

Many scientists, busily at work at their bench, will be surprised to hear this. But throughout history every great society has been subject to the dispute of competing parties under three headings: power, production, and belief. Science, far from merely being a joyous activity within the walls of the laboratory, has been more deeply involved in all three than almost any other pursuit. Since the early seventeenth century, the sciences have more and more aggressively asserted their primacy under each of these headings, at the cost of the previous occupants. Since Francis Bacon and Isaac Newton, who respectively promised omnipotence and omniscience, and whose followers have continued to brandish these hopes, science and science-driven technology have worked hard to penetrate into and transform this whole triad of power, production, and belief. It was not on the better calculation of planetary orbits or cannon ball trajectories that scientists in the seventeenth century based their chief claim to attention, but on their role in replacing the whole pre-scientific belief system. For over three centuries since then, they have pointed to their grand program of fashioning an irresistible, overarching, well-integrated world conception based on rational science. Of course such an imperious project scandalized the previous chief cultural dominators of Western society, and they have resisted being nudged aside.

During the nineteenth century, the claim of science became secularized, but otherwise only increased in its ambitions. James Frazer, author of the *Golden Bough,* taught that Western civilization has passed successively in stages from myth to religion to science. Of course he was wrong—we exist today still in a boiling mixture of these three systems, and the mutual challenges and attempts to delegitimate one another as the foundation of our culture have continued. Thus the nineteenth-century Romantics wanted to put what they called their Visionary Physics in place of the mechanistic one of their day, holding with the poet Blake that Newton, Locke, and Bacon were the "infernal Trinity" that had satanic influences on humanity. Parallel to these beliefs, that century saw a flourishing of mesmerism, phrenology, table-raising spiritualism, and the electrical creation of life forms.

Today there exist a number of different groups which from their various perspectives oppose what they conceive of as the hegemony of science-as-done-today in our culture. These groups do not form a coherent movement, and indeed have little interest in one another; some focus on the epistemological claims of science, others on its effects via technology, others still long for a return to a romanticized pre-modern version of science. But what they do have in common is that each, in its own way, advocates nothing less than the end of science as we know it. That is what makes these disparate assemblages operationally members of a loose consortium.

The most prominent portions of this current counter-constituency, this cohort of delegitimators, are four in number. Starting from the intellectually most serious end, there is a type of modern philosopher who asserts that science can now claim no more than the status of one of the "social myths"—the term used by Mary Hesse[9]—not to speak of a new wing of sociologists of science who wish, in Bruno Latour's words, to "abolish the distinction between science and fiction."[10]

Next, there is a group, small but very influential, of alienated intellectuals, of whom Arthur Koestler served as prominent exemplar. For them to be doomed to ignorance is the worst wound. But the fantastic growth rate of new knowledge and our spotty record as educators have left them impotent, and, as Lionel Trilling honorably confessed, inflict on them a devastating "humiliation."[11] In this way, powerful intellectuals who in previous centuries would have been among the friends and most useful critics of science (as the more thoughtful cultural critics still are) find themselves abandoned—and in exasperation write attacks on science such as are to be found in Koestler's later books.

Third, there is a resurgence among what I have called the Dionysians, with their dedication ranging from New Age thinking to wishful parallelism with Eastern mysticism.[12] Some have their roots in nineteenth-century Romanticism, some in the 1960s' countercultures; but all agree that one of the worst sins of modern thought is the concept of objectively reachable data.

A fourth group, again very different, is a radical wing of the movement represented by such writers as Sandra Harding, who claims that physics today "is a poor model [even] for physics itself."[13] For her, science now has the fatal flaw of "androcentrism"; that, together with faith in the progressiveness of scientific rationality, has brought us to the point where, she writes, "a more radical intellectual, moral, social, and political revolution [is called for] than the founders of modern Western cultures could have imagined."[14] One of her like-minded colleagues goes even further, into the fantasy that science is the projection of Oedipal obsessions with such notions as force, energy, power, or conflict.

That these groups have been able to gain considerable attention is due in part to the fact that the ground for dismay with modern science and technology has been prepared by three different factors, all operating in the same direction. Two are international in character, the third is local to the United States, and all of them play into the hands of the intentional delegitimators.

First, with science and engineering now central components of modern life, from birth to death, it is not surprising that concern is widespread over some real or imagined consequences of science-driven technology, nor that some of these have in fact been first examined and made public by scientists and engineers. Interestingly, we hear less now about a feared displacement of human labor by machines, which agitated the United States during the Great Depression. The concern today is closer to that expressed by Franklin D. Roosevelt in his second inaugural address and again later in 1937 in a letter to President K. T. Compton of the Massachusetts Institute of Technology, where Roosevelt wrote that the engineer's responsibility should include considering "social processes," "more perfect adjustment to environment," and designing mechanisms "to absorb the shocks of the impact of science."[15] Today's interrogators of engineering tend to go much further, fearing that technological devices, if mismanaged, can lead to the technologization of barbarism or curtail the life-sustaining capacity of this globe. The ordinary "man in the street" who harbors such fears is not convinced that the bulk of the community of scientists and engineers is sufficiently dedicated to the containment of these threats or that its protests are taken seriously at high policy levels.

This leads us to the second factor, of which the now international ecology movement is an indicator. Earlier than even most scientists, some critics intuited the fragility and delicacy of the interconnections that govern the well-being of all species on Earth. Their methodology and their rhetoric may not always have been sound, but their motivation has been a Darwinian one.

The need for ecological-systems thinking, both for its benign significance and because of the evident threats, is rather new, having emerged into global thought only in the last third of the twentieth century, and is bound to become a chief preoccupation of the twenty-first. There were of course very significant pioneers earlier, such as John Muir and Patrick Geddes, who prepared our minds in terms of their local or localizable concerns. Even Rachel Carson was focusing only on the threats to the ecosystem from certain chemicals. We now treasure these pioneers even more, because they prepared us to understand better the global meaning that had to be extrapolated from their messages. We now know that a relatively local insult to the ecosystem can and often does produce effects far "downwind." One thinks here of the discovery of dangerous radioactive fallout from A-bomb tests; the disaster for Indian farmers traceable to deforestation in Nepal; the effect on people and agriculture owing to the Chernobyl disaster; the tragedy of the decimation of the Amazonian rain forest; the widespread pollution at the Hanford Engineer Works, along the Rhine, at Love Canal, and many other sites; the linkage of droughts and floods to poor land management far away; and of course, insistently—as if the globe as a whole were rousing itself to shout for our attention—ozone depletion and the greenhouse effect. Once again, citizens who are reaching out for a new ethos of global stewardship find that they have relatively few visible and vocal allies among the academic scientists and engineers, even fewer among their brethren in industry.

Last but not least, with the rise of many scientists to prominence in our own nation's life, something was triggered in the American response which is perhaps idiosyncratic for this country but in fact is fundamentally healthy—namely, skepticism against this, as against any, form of strong, organized authority. As the astute political scientist Don K. Price has pointed out, Americans tend to have a special response to science, one that has roots in our ingrained political philosophy. From the beginning, the predominant attitude toward any large-scale organized authority in the United States has been essentially negative, and our political institutions are set up with the purpose of impeding the assertion of centralized authority as far as possible. In the first century and a half of the Republic, scientists and engineers were seen as outsiders, even as a force against established authority, as challengers of all dogma and successors of the religious dissenters who founded this country. When Joseph Priestley, equally unorthodox as a chemist, political writer, and theologian, fled England and the mob that destroyed his house, library, and laboratory, Thomas Jefferson embraced him on his arrival to America as a fellow dissident against the King and his Church. Scientists became the inheritors of the belief in progress.

But, Price says, "during the past generation there has been a sharp break with this tradition."[16] As scientists have become far more numerous and their work, directly or indirectly, has begun to change our daily lives, they have come to be identified not with dissent but with authority. Thus, although science itself is still seen as a positive force by a majority of Americans, scientists—who have been slow to understand this reaction—have become increasingly targets of suspicion.

A Framework for the Notion of *Weltbild*

Thus the forces of delegitimation of conventional science and of its claims have been receiving powerful assistance from various recent historical developments. Our next level of scrutiny of the multi-faceted anti-science phenomenon depends on using a set of ideas and postulates of which I can provide here only a sketch of the analytic base. Put in terms of a topical outline, it runs as follows.

1. While the actions of individuals, taken in the midst of practical and conflicting social realities, cannot be explained in simple terms, studies in anthropology, psychology, sociology, the history of science, and other fields show us that opinions and actions are to some degree guided by *a generally robust, map-like constellation of the individual's underlying beliefs of how the world as a whole operates.* It is a representation of reality in which, as Max Weber noted, "events are not just there and happening, but they have meaning, and happen because of that meaning."[17] For example: research on personal values inventories in the United States has shown that in individual cases one can identify constellations of important primary components of a general world picture, such as patriotism + religion + national security + stability + "morality," one that could be summarized by the term *traditionalism.*

2. The constellation of underlying beliefs forming the individual mind map is *not necessarily internally coherent or noncontradictory.* On the contrary, it is quite likely to have internal contradictions, may even harbor grotesque excesses, and yet tend to resist disconfirmation. Examples would include practicing slavery and believing "all men are created equal"; and in Nazi Germany the "purification" of science with the active participation of German scientists, not to speak of well-trained German doctors seeing themselves as social "healers" while participating in extermination.

3. The individual set of basic beliefs is *not necessarily stable over time.* There can be significant changes; persons can sometimes pass through the barriers between radically different belief systems. (When young, Empress Catherine II of Russia considered herself a

devoted "pupil" of Voltaire and his Enlightenment outlook; later, she angrily relegated his portrait bust to the attic.)

4. The constellation of a person's underlying beliefs, the belief system organized in an ideational world, has long ago received the useful, but now (at least in the English language) somewhat debased, term *world view,* or *world picture,* or *Weltbild.* This concept has considerable overlap with Robert K. Merton's important notion of *sentiments* or of "an emotionally consistent circle of sentiments and beliefs"; these support and express themselves in word or deed, which in its turn "reacts upon the sentiments, re-enforcing, moulding, at times altering them so that the whole process is one of incessant interaction."[18]

5. The world pictures of any two individuals may be largely (if only temporarily) *compatible, or in conflict, or "orthogonal" to each other.*

6. At any given time in a given culture, *many clusters of partly overlapping individual world pictures* will be discernible (e.g., "environmentalists" vs. "high frontier enthusiasts," or "traditionalists" vs. "individualists," or "family orientation" vs. "achievement orientation"); but one can sometimes discern, perhaps more clearly in retrospect than contemporaneously, among those competing clusters a *dominant* world picture that may characterize an era, or at least a *dominant set of components* among the current world pictures.

7. Because personal variants occur within a big envelope, and each world picture has many components, *no single variant is entitled to be considered the "pure" case.*

8. Each of the various individual world pictures, including their scientific core, is *internally functional in its own terms,* although from the point of view of a different world picture it may be regarded as unsuitable. Thus, a navigator still properly calculates positions by means of a geocentric model. The Zinecanteco Maya of Mexico have a satisfactory theory of earthquakes in terms of the sudden movements by the four giants on whose shoulders the corners of the cubical Earth are said to rest. Similarly, the scientific ideas of undereducated but generally strongly "pro-science" children and of "scientifically illiterate" adults form a complex but functional *science sauvage.*[19] In *science sauvage,* the facts of nature form a seemingly infinite, atomistic, unconnected set; material bodies come to

a stop unless they continue to be propelled; electricity flows through wires as water does through pipes, only much faster; space is a big container in which matter appeared at the beginning of time; time is everywhere the same and marches on inexorably on its own; notions of probability and scaling are minimal; science and engineering are hardly distinguishable; the pattern of cause and effect works most of the time, but unfathomable and magical things do occasionally intervene; science provides truths, but now and then everything previously known turns out to have been entirely wrong, and a revolution is needed to establish the real truth. And so forth.

9. *No world picture is truly anti-scientific,* insofar as it always has a core component containing a functional proto-theory of the physical and biological universe.

10. A basic function of a world picture is that it *acts as a cohesive force for the formation and work of a community.* As Erik Erikson put it, "A world view, then, is an all-inclusive conception which, when it is historically viable, integrates a group's imagery. According to our formula, it focuses disciplined attention on a selection of verifiable facts; it liberates a joint vision which enhances a sense of historical reality; and it actualizes a widening fellowship with strong work commitments. Altogether, these tendencies seem to confirm some historical truth previously only intuited." In this way, Erikson noted, the individual "can boast of a sense of centrality of the world and leeway in action."[20] And just as for scientists or any other group, the world view which incorporates the countervision also provides a sense of community and common action.

11. Since by definition a world picture is a system that helps us understand how the world as a whole operates, and that acts as a cohesive force for community formation, it can (and often does) *exclude the individual's private, personal, imaginative parts of experience,* such as one's interaction with the arts; the purposes of the world picture are primarily public, social, and epistemological.

12. Both an individual's and a community's world picture *define themselves with respect to their opponent(s) or opposite(s).* This fundamental fact—true even for the words of a language (note the original structure of the *Thesaurus*)—is familiar to historians of science, who have learned that over the past centuries the so-called

scientific world pictures arose in terms of demarcation from their opposites, usually their predecessors (e.g., the succession Newtonian, Romantic, mechanistic, electromagnetic, relativistic, empiricist world pictures).

13. The *scientific* world picture, whether "adequate" or "modern" or not, is an *embedded portion* within an individual's general world picture.

14. There is usually *some plausible connection* among the general and the scientific / technological components of a world picture. (It is not just a quirk that caused Kaiser Franz Joseph to object to the use of the automobile, the telephone, and indoor plumbing; they did not fit into his tradition-bound mind map. Conversely, Peter Galison has shown that one can find coherence in one modern world picture of science, epistemology, architecture, and liberal politics.)[21]

15. However, *there need not be such coherence in every case.* (Note, for example, the initially surprising tendency of the frequently encountered constellation of religious fundamentalism + creationism + high regard for technology. But science can be used strategically by anti-science movements that have political goals.)

16. For the population at large, favoring astrology, mysticism, faith-healing, and the like are attitudes that *form only the surface phenomena,* or by-products, of a world view. They are nourished by the more fundamental set of beliefs in a particular *Weltbild.* But so are the "pro-scientific" attitudes.

17. If a world picture that does not include as one of its components the standard Western scientific world picture, it is likely to be *perceived as a counter–world picture* by those whose world picture does include it.

18. But the situation is *symmetrical:* any one of the two can be taken as a counter–world picture of the other.

19. It might be more appropriate to use the term *alternative science* than *anti-science,* except that the word *alternative* allows the impression that such conceptions are on the same ontological or pragmatic level as "real" science. Hence, one may prefer the term *parascience.*

20. At the core of any world picture, as its major cognitive structure of epistemological significance, is *a set of thematic concepts and*

presuppositions; these are largely unconsciously held, untestable, quasi-axiomatic ground beliefs that have been found functional by its devotees. In the specific case of scientific world pictures, examples of these thematic hypotheses or thematic propositions have at different times been explanatory schemes based on the thema "hierarchy" or its anti-thematic opposite, "holism"; mechanistic vs. mathematical models; favoring of vitalism vs. materialism; evolution vs. steady state vs. devolution.[22] By contrast, in a religiously based world picture, the "themata of relation," according to Gerhart and Russell, are "to begin with obviously . . . what have been called the traditional doctrines: God, grace, sin. Stated more philosophically, for example by Kant, they are God, freedom, immortality; or more contemporarily, the sacred, world, humankind. These three themata, variously expressed, are a representative but not exhaustive list."[23]

21. What distinguishes a world picture most basically from its opposite, alternative world picture is *the incorporation of a significant number of anti-themata* in the second, in place of the themata of the first.

22. A *Weltbild* and its counter-*Weltbild* may be mutually *incompatible, but they are not logically incommensurable.* (For example, environmentalists such as Carl Sagan and technology enthusiasts such as Edward Teller have generally not been separated by mutual misunderstandings about their concepts or aims.)

23. Just as the individual's world conception can be subject to change in time, the allegiance of a group to a particular set of themata in a world picture can be *time-dependent;* that is to say, the hold which some themata have on a community may decrease in intensity in one world picture, while their opposites in another community gain. In this way, a seemingly "new" predominant world picture may take center stage over time. (For example, Pope John Paul II has accepted many of the claims rejected in the trial of Galileo, including the equal standing and authority of scientific findings.)

24. Both for individuals and for a community, changes in allegiances to a particular set of themata in a world picture can sometimes be seen to be *correlated with changes in external* (e.g., political, economic) *conditions* that test or challenge the functionality of the

existing general world picture. (For example, wider acceptance of Copernicanism in the aftermath of New World explorations; rise of scientific materialism after the political turmoil of the 1840s, and of positivism after the collapse of the Hapsburg Empire; efflorescence of anti-science movements in the United States during the Depression and during the Vietnam War; Lewis Mumford's switch from early technology scholar to 1960s-style anti-science critic.)

25. In addition, changes of thematic allegiances within a given world picture are likely to *lay bare or exacerbate pre-existing internal contradictions or conflicts.* (For example, the rise of counter-scientific movements embracing holism in post–World War I Germany; short-term popular enthusiasm for nuclear "victory" and for science in 1945 vs. long-term reassessment of these in the post-Hiroshima era; the reappearance of latent ethnic, religious, and regional rivalries in the wake of the dismantling of the State ideology in Eastern Europe.)

26. The increase in awareness of internal contradictions in a world picture, brought about by external stress, can provide the *opportunity for the most effective educating intervention* to take place. (Positive examples are the strategies of Gandhi and Martin Luther King.) Such intervention, rather than simply trying to "correct ignorance" or "dislodge error," is the most promising way to deal with dysfunctions, including symptoms of the devaluation and delegitimation of conventional science. An attitude toward science that contains inherent flaws may not be immune to even a relatively brief intervention. A revealing example came in a pilot experiment initiated in 1980 with The Public Agenda Foundation. In it, six population-representative groups of 9–14 persons each in different cities throughout the U.S. were convened for extensive discussion to decide on questions of policy or ethics in which large components of scientific and technical understanding seemed to be required (e.g., the wisdom of increasing aggressive research on aging, on separation of isotopes of fissionable material, etc.). At the beginning of the evening, each of the several groups that participated produced a rather predictable "top-of-the-head" response that revealed the usual level of scientific-technological ignorance, as found in many polls. But at the end of the evening, after the group had been forced to debate the scientific and technical aspects with

the aid of explanatory materials made available to it and had wrangled with one another, a second vote was taken on the same matter. As it turned out, the result of the second vote ("considered judgments") was quite different from the first and approached the results obtained separately from groups of professional scientists considering the same questions. Thus, with some care and resources, one can hope to "work through" problems of science and technology that have social and political dimensions, even in a relatively brief period and with apparently ill-prepared groups.[24]

27. The predominant world picture of a society or community at a given time is difficult to understand unless one has studied what have been called *the specific historic "particularisms" or "exceptionalisms"* of that society or community. This is certainly true for America, the more so as America's exceptionalisms, from the colonial beginnings, have been closely connected with ideas of science and its social organization. Basic local facts that puzzle foreign observers, such as a desire of the Founders to have the Constitution reflect Newtonian physics and cosmology, and yet the absence to this day of anything like a centralized Ministry of Science, greatly influence what science means to Americans.[25] A parallel exceptionalism was operating in the Soviet Union, for it too (in its own way, and with very different outcomes) at the founding of the State based its original hopes on lessons said to be drawn from science.

28. Finally, in an individual's or community's general world picture *the scientific and the political components tend to seek mutual accommodation and reinforcement.* This can result in greater coherence when each of the components is well structured; but it can also result in greater lability for the whole if these parts are themselves each in disarray and thus raise the potential for sudden and catastrophic shifts of the total world view.[26]

What Is Modernity? A Sociological View

Having sketched out a framework to help us to think seriously about confrontations between supporters of science and anti-science, we can now turn to seek out the subterranean connections between uninformed or hostile attitudes to science and the general world pictures of which they are an expression. Here we encounter immediately the hypothesis that the hostility to or disinterest in the scientific view chiefly indicates opposition to a world view which we might identify by the term *modern.*

That does not imply that *modern* is necessarily equivalent to *better*—some very intelligent persons today are not enamored of modernism[27]—nor can one deny that heads have rolled over the definition of this troublesome concept, not least over the demarcation between modern and contemporary in art history. Yet, we must at least point to where one may find it in an operational manner.

At any given time, in the competition between the currently widely accepted general world picture and its various opponents, there is an area where the challenge is intense, a moving interface of contemporaneity at which there is a state of potential or real confrontation. To use the usual but inadequate terms, that is the arena for distinguishing between the "traditional" and the "modern" as well as between the "modern" and the "post-modern." The experience of this sometimes passionate encounter is known to each of us in personal experience. For some older people today, the daring embrace in their youth of a new vision beyond the then reigning modernism was perhaps exemplified by championing against all odds Freud, Stravinsky, Brecht, Gropius, Joyce . . . together with a liberal theory of history that ran from John Locke to Bertrand Russell; and they see, as the frontier has moved, that a new generation now finds much of this to be antiquated detritus, being nudged aside by the new anti-canon (Lacan? John Cage? Robert Wilson? etc.).

In truth, modernism is a protean concept, assuming ever-changing aspects. In Leszek Kolakowski's phrase, modernism has been on endless trial. At the time when Galileo proposed the set of four great novelties that subsequently became part of our "modern" scientific world picture—the quantification of nature, the mechanization of nature, the distancing of the world of direct daily experience from the world of science, and last but not least, secularization—the then established outlook in Italy, exemplified by the sophisticated work of the Jesuit scientists, found itself diametrically opposed by the Galilean set. If the terminology had existed then, the Jesuits might have called themselves modern, and Galileo postmodern. Similarly, the Newtonian world was as strange to the ordinary eighteenth-century person as quarks and ten-dimensionality are to today's nonscientists.

There is little doubt that the everyday "modern" person, with all his or her faults, was slow to come to the fore in the West. We get an early glimpse of modern man, for example, in Gustave Flaubert's meticulous attempt at realist rendition in his novel *Madame Bovary*—a work that was too advanced for the 1850s and resulted in Flaubert's prosecution. Flaubert's modern man—the only character in the book who survives the general catastrophe that sweeps all others and their world away—is the second-rate, insufferable pharmacist, Monsieur Homais, who confesses to the religion of "the God of Socrates, of Franklin, of Voltaire . . . and the immortal principles of '89." At the end it is only he and his family who are "flourishing and merry . . . with whom everything was prospering." His sons, named Napoleon and Franklin, "helped him in the laboratory . . . and recited Pythagoras' table in a breath." Assiduously he sends off his observations on the manufacture of cider and the behavior of the plant louse to the scientific academies. And it is to the ascent of this new person that the last sentence of the novel is dedicated: "He has just received the Cross of the Legion of Honor."

But for our purposes we need not enter into every aspect of the debate about what modernism means and when it began. To define properly the mind map of the modern person at our phase of history, one would indeed have to triangulate from several different points. For us it will be sufficient if we arrive at an operational notion of *modern* by seeking the intersection of sightlines that start from only two bases, one being sociology, the other the history of ideas. We shall now briefly look at the results obtained from each—and will find that these results converge.

An example of the sociological approach is the pioneering work of Alex Inkeles and his associates at Stanford University, starting with *Becoming Modern.*[28] There are other candidates, but for our purposes the Inkeles group's findings will be a good start.[29] By examining 1,000 persons in each of six "developing" countries, from Chile to Israel and India, the researchers sought cross-national, trans-cultural results that would reveal "not only a potential but . . . an actual psychic unity in mankind." Thereby a person who is modern in one culture would be recognizable as modern in another, apart from "the distinctive attitudes with which his culture may otherwise have endowed him" (p. 118).

Four basic criteria emerged that define someone as modern in our time trans-culturally: being an informed participant citizen; having a marked sense of personal efficacy (feeling able to control one's own destiny and events in the world); being highly independent and autonomous; and being open to new ideas and experiences ("cognitively flexible"), including in particular exhibiting interest in technical innovations and in the scientific exploration of previously sacred or taboo subjects. Such traits are needed and reinforced by, and in turn need and reinforce, some of the obvious characteristics of modern institutions (e.g., the factory) which "need individuals who can keep to fixed schedules, observe abstract rules, make judgments on the basis of objective evidence, and follow authorities legitimated not by traditional or religious sanctions but by technical competence" (p. 4). As one would expect if one holds that industrialization and bureaucratization tend to reorganize and rationalize all aspects of life, the modern personality identified in this study matches the exemplar of the modern urban-based, industrial order that demands the acceptance of conditions which the authors define as follows: personal mobility; readiness to adapt to changes in working and living; an innovative as well as utilitarian spirit; and tolerance of impersonality, of impartiality, and of differences among people. These contrast with the tribal or old order, characterized by passivity, preference for the status quo, and subordination of the individual self to higher authority.

Among the analytic criteria also used in those studies, our eye is caught by such criteria as the amount of information possessed on various topics; orientation to the present or future rather than the past; valuing of technical skill and education; belief in the possibility of human control over the social and natural environment; long-term planning; and the value of science as such, particularly the belief in the calculability, predictability, and causal lawfulness of the physical-biological world.

Not surprisingly, and in line with our stress on the interconnectedness of elements within the general and the scientific parts of a world picture, we find that the modern person as defined by the criteria given above also exhibits characteristic behavior and opinions under such headings as kinship and family (threats to the wider family due to mobility, etc., but strengthening of immediate family ties), women's rights (including favoring of birth control), religion (increase of secularism), politics (desire for participation), and social stratification (status connected with skill and education).

The trans-cultural psychosocial portrait of this "modern person," derived from studies in developing countries, paints a rather coherent, internally articulated world picture, even if it is applicable to only a minority within the general population. (Moreover, for reasons given by the authors, such as the concentration of men in individual jobs at that time in those countries, the samples studied contained apparently only men, and the authors make plain that this is a source of concern and a call for future additional work. They also point to preliminary evidence that "the pattern which will eventually emerge for women will be broadly similar to what we observed in the case of men" (p. 311). I am not aware of a similar empirical study of the incidence of modernity having been conducted in the supposedly more "modern" population in the United States or Europe; but there, too, a substantial fraction undoubtedly either falls short of matching even moderately well the portrait of exemplary modernity drawn here or at least carries self-contradictory attitudes side by side.[30]

The most obvious anti-modern characteristic of that fraction among our population is precisely its embrace, within its world picture, of parascience in its various forms, from astrology to "enchanted" science—elements that necessarily contradict the criteria of modernity cited above, such as tolerance of impersonality and the valuing of (conventional) science as such, particularly its belief in the calculability, predictability, and causal lawfulness of the physical-biological world.

Consider, for example, the criterion of tolerance of impersonality. Within the scientific segment of the modern world view, it is an essential point. Perhaps the most basic aspect of the scientific method is that despite all personal passion and ecstasy in the doing of it, the results are to be completely invariant with respect to private longings or individual differences. Thus in his "Autobiographical Notes," Einstein spoke of his "attempt to free myself from the chains of the merely personal . . . The mental grasp of this extrapersonal world swam as highest aim . . . before my mind's eye."[31] And Max Planck, apologizing for having introduced quantization into physics, said he was above all motivated by a search for "absolutes," that is, for knowledge valid not only for all people but even, if they existed, for extra-terrestrial beings.[32] The tolerance of impersonality is at the very heart of conventional science; but it is anathema from the point of view of parascience, with its celebratory focus on the personal, its introduction of "consciousness" even into the atom, and other quasi-animistic beliefs.

Entreact: Revisiting Astrology Briefly

Letting a belief in astrology—in itself usually harmless—stand for the moment for the whole complex of parascience, we may pause here to underline the points just made by a fortunate example, furnished in an essay by the novelist (and former engineer) Kurt Vonnegut.[33] Under the cover of humor he revealed the vast gap between our list of characteristics of modernism on the one hand and the yearnings underlying parascience on the other. The occasion was a satiric and eloquent address calling for nothing less than an end to science, which Vonnegut delivered some years ago to a graduating class at Bennington College. In his speech, he said:

> We would be a lot safer if the government would take its money out of science and put it into astrology and the reading of palms. I used to think that science would save us, and science certainly tried. But we can't stand any more tremendous explosions, either for or against democracy. Only in superstition is there hope. If you want to become a friend of civilization, then become an enemy of truth, and a fanatic for harmless balderdash . . . I beg you to believe in the most ridiculous superstition of them all: that humanity is at the center of the universe, the fulfiller and the frustrator of the grandest dreams of God Almighty.
>
> About astrology and palmistry: they are good because they make people feel vivid and full of possibilities. They are commu-

> nism at its best. Everybody has a birthday and almost everybody has a palm. Take a seemingly drab person born on August 3rd, for instance. He is a Leo. He is proud, generous, trusting, energetic, domineering, and authoritative! All Leos are! He is ruled by the Sun! His gems are the ruby and diamond! His color is orange! His metal is gold! This is a *nobody?* . . . Ask him to show you his amazing palms. What a fantastic heart line he has! Be on your guard, girls. Have you ever seen a Hill of the Moon like this? Wow! This is some human being!

Vonnegut ended his implied case against science by extolling the arts, whose purpose, he said,

> in common with astrology, is to use frauds in order to make human beings seem more wonderful than they really are. Dancers show us human beings who move much more gracefully than human beings really move . . . Singers and musicians show us human beings making sounds far more lovely than human beings really make . . . And on and on. The arts put man at the center of the universe, whether he belongs there or not.

But science, he says, fails to do that; and "military science . . . treats man as garbage—and his children, and his cities, too."

Modernity: A Philosopher's View

We shall return to this pregnant text shortly. But first we need to complete the characterization of modernism that was promised, now from a point of view different from sociology, namely from intellectual history and philosophy. Here the time base shifts somewhat. Unlike the earlier account that defined modernism directly in twentieth-century terms, modernism is here more likely to be considered as the continuing inheritance of a transition from humanism to rationalism. From the vast literature I select one author, precisely because he is not unsympathetic to the anti-science phenomenon. He takes his stand somewhere in the middle between the extremes, marked at one end, say, by Morris Berman in his *The Reenchantment of the World*[34] and at the other end by the earnest remnants of the Vienna Circle positivists, descending from the manifesto "Wissenschaftliche Weltauffassung," whom we discussed in Chapter 1. The author I have chosen is the philosopher Stephen Toulmin. In his book *Cosmopolis* he tries to find—to use the subtitle of the work—"the hidden agenda of modernity."[35] In a frankly speculative but generally sober manner, he identifies in intellectual history the rise of the principal elements of post-Cartesian modernism, which he terms "High Modernity."

The "timbers of the Modern Framework" are of two kinds and concern, respectively, nature and humanity. As to the first, modernism (as Toulmin defines it) is characterized by beliefs such as these: "Nature is governed by fixed laws set up at creation . . . The objects of physical nature are composed of inert matter; so, physical objects and processes do not think," etc. (p. 109). As to the second, we find: "The 'human' thing about humanity is its capacity for rational thought or action; rationality and causality follow different rules; . . . so human beings live mixed lives, part rational and part causal . . . Emotion typically frustrates and distorts the work of reason," etc.

But this, Toulmin holds, is not the unchallenged state of affairs any longer. The timbers of the post-Cartesian framework have come to be gradually dismantled, particularly during this century, giving way in our time to what he terms "Humanism Reinvented" at the far side of modernity. Across the whole range, from the centrality of inert matter to the separation of reason from emotion, the twentieth-century scientists themselves furnished ammunition against these doctrines. They have moved from the historical, concrete, and psychological toward the formal, abstract, and logical; from the search for overarching certainties and unification of knowledge to the acceptance of specific indeterminacies and a confederation of equal sciences. The anti-modern movement of today, in this analysis, is at bottom a revival of Renaissance humanism, with its tolerance of uncertainty, ambiguity, and diversity, with its lack of rigor and Montaignean skepticism: it is a movement "for a reintegration of humanity with nature, a restoration of respect for Eros and the emotions, for effective trans-national institutions [after "thirty years of slaughter in the name of nationalism"], . . . an acceptance of pluralism in the sciences, and a final renunciation of philosophical fundamentalism and the Quest for Certainty" (p. 159).

The counter-culture arising in the 1960s—representing a portion of what we have been calling here the anti-science phenomenon—is therefore not seen as merely the transient effect of youth culture, nor only the response at that time to the Vietnam war. Rather it is an indicator of a dissolution of a three-centuries-long reign of a now "moribund world view," an attempt to restore unities that were dichotomized in the seventeenth century, such as "humanity versus nature, mental activity versus its material correlates, human rationality versus emotional spring of action, and so on . . . After three hundred years, we are back close to our starting point" (pp. 161, 167).

As to science today, as long as it remains rooted in experience it can now shake off any presuppositions that limit speculation: "We are freed from the exclusively theoretical agenda of rationalism" (p. 168). In this view, rationalism turns out to have been, as Heidegger indicated even in the original title of his essay on the modern *Weltbild,* nothing more than a *Holzweg*—a treacherous way that loses itself into the woods. We must make do without the dreamed-of set of uniquely authoritative principles as a basis for human knowledge, just as we must now also do without a universalistic theory of ethics or of politics.

But, Toulmin continues, this does not mean that we are now condemned to return to the world picture against which Descartes and Galileo fought, nor that we must accept a "Farewell to Reason," nor even that we must slip into that vague and chaotic condition called "post-modernism." The choice before us is not between rationality and absurdity, nor between rationalism and chaos. On the contrary, Toulmin proposes that the removal of the scaffolding of modernity allows modernity itself to "come of age," to attain a new phase in the sense of absorbing now into its agenda emancipatory ideas and commitments to egalitarian practice (which Jürgen Habermas, from his own point of view on modernism, would in fact have termed the key ideas of modernization). One gathers that this would imply, for example, a redirection of some scientific researches so as to connect them organically with the major problems besetting the human race, as discussed in Chapter 4 on the "Jeffersonian" type of research program.

Components of the Modern World Picture, and of Its Alternative

Those two analyses of the notion of modernity, proceeding from two very different bases, seem at first glance not to have a great deal in common. Our sociologist regards modernity as the preferred final stage, reached in our time, of a benign social development embedded in praxis, by which any citizen of our century, in any country, can hope to shake off the vestiges of a feudal past marked by powerlessness, superstition, and ignorance. Our philosopher, on the other hand, looks at the theoretical concept of modernism, finds it to have reached its zenith perhaps two centuries ago, and declares it to be now decaying from the High Modern peak because of the intellectual insufficiency of the supporting girders. The first thinks of Western science still as a solid mainstay within an operational world picture, while the second is ready to speculate on the rise of alternative models as part of the development of a new phase of modernism. The first thinks primarily of modernism as the ground gained with respect to the earlier phase of our condition, the second as ground being lost to a "new phase" that is trying to establish itself.

Beyond these differences, however, we discern a large overlap between them, if we keep in mind their respective time frames and preferred directions of view. Whether operational or decaying, the respective world conceptions entitled to the term *modern* contain many of the same components—and any anti-science movement would be in conflict with them in both cases. Therefore we may now make a list of the main components, traits, or tendencies of what both these commentators would regard, for better or worse, as roughly characteristic of the predominant contemporary "modern world picture" with its strong science-oriented component—the conception now being besieged on one side by the traditionalists and on the other by the self-declared postmodernists. In what follows we shall allow for the facts that individual variants occur within a big envelope and that few people would embrace every component on the list with equal dedication. We also must remember here that the world picture, by definition, is the system constructed to deal with the public, social, and epistemological parts of an individual's experience rather than with the private, personal, and imaginative ones. Thus one must not expect the components of the world picture enumerated below to tell about or interfere with one's personal aesthetic responses to the arts, or indeed the kind of transcendence possible while in the thrall of a scientific discovery (as Einstein explained in the passages given in Chapter 4).

Such a list, then, of items characterizing a modern world picture, encoded in a set of telegraphic phrases, would run as follows:

High place for "objectivity"
Preferably quantitative rather than qualitative results
Extra-personalized, universalized results, where available
Anti-individualism
Intellectualized, abstract, divorced from the sensual world of direct experience (contrary to Mach), de-eroticized, de-anthropomorphized
Rationality rather than moralistic thinking (where rationality is operationally defined by such boundary conditions as skepticism and consensuality)
Problem-oriented (versus mystery-oriented; versus purpose-oriented)
Proof-oriented (demanding verification or test of falsification)
Tendency to meritocratic functionality; "reason and routine"; specialization
Skepticism with respect to authority; autonomy-seeking
Rationalistic, Enlightenment-based, opposing sacralization of any subject
Tendency to accommodate contrary view only if proven, but open to debate and new experience (what J. Bronowski termed "democracy of intellect" instead of "aristocracy of intellect")
Scientific knowledge leads to power (e.g., Vannevar Bush's promises in *Science, the Endless Frontier* [1945])
Hierarchy exists among fields of knowledge, with the more

fundamental ones serving as sources of explanation for the rest

Avowedly secular, anti-metaphysical, "disenchanted" (F. C. S. Schiller's *Entgötterung der Natur*)

Evolutionary rather than preferring either stasis or discontinuous ("revolutionary") change

Preferably un-self-conscious, non-self-reflexive

Cosmopolitan and globalist

Active, progressive (i.e., scientific progress → material progress → moral progress, as in the evolution of human rights [contrary to Descartes])

Many of these traits have fairly obvious connections with one another and so form a robust network. Moreover, many can be elaborated to exhibit the thematic notions behind them, and the set as a whole is so close to the world view expressed by the empiricists whom we met in Chapter 1 that they might with some justice regard it as part of their long-range heritage. But a main point for us now is that this list immediately suggests how to obtain a sketch of an *alternative* though equally functional and internally coherent world picture—"pre-modern" in Inkeles's terms, or "post–High Modern" in Toulmin's. We need only remember that world pictures can be defined in terms of their opposites. That is, one only has to make a second list, line by line, of each of the corresponding antithetical tendencies; thereby one obtains almost automatically the main outline of the counter–world picture, one that would dismiss the list above as mere "scientism." It will also become clear at a glance that the so-called science implied in such a counter-construction will as a consequence have to be as different as astrology is from astronomy. Kurt Vonnegut's tongue-in-cheek text, given earlier, has served to prepare us for this finding, having been in essence a plea against the first list and in favor of the second list.

The set of telegraphic phrases characterizing the countervision (again, an idealized one) would now run as follows:

Subjective, not objective

Preferably qualitative rather than quantitative

Personalized, not extra-personalized

Ego-centered

Sensualistic and concrete, not intellectualized and abstract
Moralistic rather than instrumental rationality
Premium on uniqueness, not generalizability
Accessible to all, not only to an elite or a meritocracy
Purpose-oriented or mystery-oriented, not problem-oriented
Low interest in tests of falsifiability
Faith-based
Tendency to systems based on individual authority rather than accommodation of equally supported contrary views[36]
Power is prior to and determines knowledge, not the other way around
No hierarchies exist among fields of knowledge; they are all essentially equally authoritative
Etc.

Science Usurping God's Throne? A Countervision Explained

This account of a constellation of beliefs is helpful for understanding some earnestly expressed opposition to the construct embodied in the earlier list. An exemplar of such an opposition is a recent address by the Czechoslovakian poet, playwright, and statesman Václav Havel before the World Economic Forum in Davos, Switzerland, published with the significant title "The End of the Modern Era."[37] It amounts to the presentation of the high points of a countervision together with a glimpse of the motivation behind it; both are presented with the eloquence one would expect of that author, and therefore excellently suited for a presentation here *in extenso.*

Looking back on a century which might well be characterized, particularly by a Central European, by the forces of brutal irrationality and bestiality, in which the fates of millions were sealed by the whims of Kaiser Wilhelm, Hitler, Stalin, and their henchmen, Havel finds the chief source of trouble to be the very opposite, namely "rational, cognitive thinking," "depersonalized objectivity," and "the cult of objectivity." The "end of Communism," he writes, has brought an end not just to the 19th and 20th centuries, but to the modern age as a whole.

The modern era has been dominated by the culminating belief, expressed in different forms, that the world—and Being as such—is a wholly knowable system governed by a finite number of universal laws that man can grasp and rationally direct for his own benefit. This era, beginning in the Renaissance and developing from the Enlightenment to socialism, from positivism to scientism, from the Industrial Revolution to the information revolution, was characterized by rapid advances in rational, cognitive thinking.

This, in turn, gave rise to the proud belief that man, as the pinnacle of everything that exists, was capable of objectively describing, explaining and controlling everything that exists, and of possessing the one and only truth about the world. It was an era in which there was a cult of depersonalized objectivity, an era in which objective knowledge was amassed and technologically exploited, an era of belief in automatic progress brokered by the scientific method. It was an era of systems, institutions, mechanisms and statistical averages. It was an era of ideologies, doctrines, interpretations of reality, an era in which the goal was to find a universal theory of the world, and thus a universal key to unlock its prosperity.

Communism was the perverse extreme of this trend . . . The fall of Communism can be regarded as a sign that modern thought—based on the premise that the world is objectively knowable, and that the knowledge so obtained can be absolutely generalized—has come to a final crisis. This era has created the first global, or planetary, technical civilization, but it has reached the limit of its potential, the point beyond which the abyss begins . . .

Traditional science, with its usual coolness, can describe the different ways we might destroy ourselves, but it cannot offer us truly effective and practicable instructions on how to avert them . . .

The world today is a world in which generality, objectivity and universality are in crisis . . . Many of the traditional mechanisms of democracy created and developed and conserved in the modern era are so linked to the cult of objectivity and statistical average that they can annul human individuality . . .

Despite Havel's further suggestion of a possible blending of the "construction of universal systemic solutions," or "scientific representation and analysis," with the authority of "personal experience," so as to achieve a "new, postmodern face" for politics, the chief animus here is of the same sort as it was in Vonnegut's disarming piece. Havel's identification of the end of the modern era is not to be understood merely as a plea for some compromise between the rival constructs; that much was made clear in an earlier version that dealt with the place of modern science quite unambiguously:[38]

> [Ours is] an epoch which denies the binding importance of personal experience—including the experience of mystery and of the absolute—and displaces the personally experienced absolute as the measure of the world with a new, man-made absolute, devoid of mystery, free of the "whims" of subjectivity and, as such, impersonal and inhuman. It is the absolute of so-called objectivity: the objective, rational cognition of the scientific model of the world.
>
> Modern science, constructing its universally valid image of the world, thus crashes through the bounds of the natural world which it can understand only as a prison of prejudices from which we must break out into the light of objectively verified truth . . . With that, of course, it abolishes as mere fiction even the innermost foundation of our natural world; it kills God and takes his place on the vacant throne, so that henceforth it would be science which would hold the order of being in its hand as its sole legitimate guardian and be the sole legitimate arbiter of all relevant truth. For after all, it is only science that rises above all individual subjective truths and replaces them with a superior, trans-subjective, trans-personal truth which is truly objective and universal.
>
> Modern rationalism and modern science, through the work of man that, as all human works, developed within our natural world, now systematically leave it behind, deny it, degrade and defame it—and, of course, at the same time colonize it.

The rhetorical power of appeals such as Havel's is strengthened by the asymmetry between the two lists above. The next step is not difficult to guess. In a variety of cases throughout history, opinions hostile to science prepared the ground for incorporating the opposition to the asserted claims of science into a larger system with room for a counter- or parascience. They include such diverse instances as Goethe's anti-Newtonianism, Blake's Visionary Physics, the "Aryan" science in Germany, the belief system of the 1960s counterculture, the anti-science campaign associated with China's Cultural Revolution, and at least one of today's cults and beliefs, as we shall see shortly.

Three Types of Ameliorating Strategies, and Their Limits

We began by asking the question whether the multi-faceted anti-science phenomenon, even if widespread, is at bottom only a more or less harmless diversion, or whether it signals an important cultural challenge and must therefore be taken seriously.

The answer is now clear. If we leave aside as comparatively unimportant the passing fads, ignorance, banalizations, and their commercial exploitation, we can focus on pseudo- or parascientific schemes that arise from deep conviction. These are grounded in a fairly stable and functional, motivating world view. It is these that can be directed at the core of contemporary culture (as would, for example, an analogous anti-literature phenomenon: in fact, some of the new cultural movements in the United States have just that purpose). Even though the counter-constructs embodying parascience are a minority view today in the United States, their entrenchment is a living reminder of an old, worldwide struggle of mutual delegitimation of rival cultural claimants. How alarming this is felt to be depends of course on one's degree of satisfaction with or allegiance to the modern world picture. And what the likely trend of this conflict may be in the near future will depend to some degree on whether earnest and successful interventions are undertaken in opposition to the counter-construct, or whether intellectuals and policy makers on the whole will continue to give only lip service to this problem, as they have done with scientific and general cultural illiteracy.

As a practical matter, there seem to be only three types of interventions that make sense:

1. The traditional one, which has now become difficult to carry out: formation, from early age on, of a modern world view that will preempt the attractions of its opposite. This implies not only early support of the child by a sound educational system designed for this purpose (e.g., with curriculum materials specifically tailored to explain the power and limits of science, such as Project 2061 of the American Association for the Advancement of Science);[39] one would also need the support by that individual's parents, teachers, and other caregivers, who themselves should have passed through an education of this kind.

2. Less intensely, and less likely to succeed on a large scale though easier to mount: interactions of the sort described above in the Public Agenda Foundation mode, that bring to light directly the internal contradictions in the alternative picture; or massive and persistent adult education efforts, such as the Open University in Great Britain (which, unfortunately, has no equivalent in the United States).

3. Still less likely to yield success but still easier in principle: widely visible exposure of the failures of the claims of parascience and persistent action to prevent its formal acceptance into schooling systems. Thus (as an example to which we shall return at the end), while convinced followers of "creationism" themselves are probably unreachable owing to the robustness and internal functionality of their supporting world picture, at least one can reverse, as was done recently in Texas after a decade-long fight, the stranglehold of these powerfully presented minority views on the selectors of textbooks for the whole state's school system.

These three normal modalities are worth further discussion, development, and implementation. But in fairness to history one must remember that there exists a process not accessible to educators that can change a central portion of the world view profoundly and quickly, particularly in those periods when turbulent external circumstances act to crack the mold. In such cases, the unexpected intellectual discontinuities and / or the new social conditions throw the reigning world picture suddenly into doubt. One thinks here of the effects of the discovery of the New World, the telescopic findings of the early seventeenth century, the great earthquake of Lisbon in 1755, the American and the French Revolutions of the eighteenth century, the crushing local hardships attending the rapid spread of the Industrial Revolution, the uprisings of 1848–49 in Europe, and the wars ending in 1918, 1945, and the 1970s. The unexpected, rather sudden termination of the Cold War and the outbreak of Glasnost may come to be seen as historic leverage points of that sort.

Some of the events just referred to helped in fact to forge components of what we now call the modern scientific world view (e.g., following the discoveries by Columbus in 1492 and Galileo in 1609; or the rise of anti-Hegelian ideas and scientific materialism in the wake of the failed revolutions of the 1840s).[40] But in most cases the world-shaking event had, at least in the short run, the opposite result, by providing audiences and respectability to the countervisions. This, in conclusion, is what we must scrutinize most attentively.

Toward a Conclusion

Among examples that help us derive guidelines from this analysis are two in particular. One is the rise of the machine-breaking Luddites in Britain from 1811 to 1816. It was a movement first spawned by economic grievances, but it eventually became a violent explosion against the technological symbols of a suffocating and unyielding factory system.[41] Here we need only refer to it, because it has a certain overlap with the other example, which took place in the 1920s and early 1930s. In the early phase of the growth of Nazism in Germany, there arose, in the words of Fritz Stern, the "cultural Luddites, who in their resentment of modernity sought to smash the whole machinery of culture."[42]

In that case, the discontent with industrial civilization joined with the reaction against aspects of the program of modernity identifiable with "the growing power of liberalism and secularism." The gathering fury did not fail to include prominently science itself. One of the German ideologues most widely read in the 1920s was Julius Langbehn, who taught that there is an opposition between the scientific and the creative and who decried science, especially its tendency to splinter into specialization. In Stern's words: "Hatred of science dominated all of Langbehn's thought . . . To Langbehn, science signified positivism, rationalism, empiricism, mechanistic materialism, technology, scepticism, dogmatism, and specialization . . ." (p. 122).

Thus it was not an accident that conventional science came under siege in Germany well before the Nazis assumed governmental power—with some German scientists demanding an "Aryan" science that was based on intuitive concepts, on the ether (as a residence of "Geist," or "Spirit"), on experimental rather than formalistic or abstract conceptions, and above all on advances "made by Germans." Spengler's conceptions seemed tailor-made to be incorporated in the Nazi ideology, and it was to his great credit that he courageously repelled all efforts to draw him into that net. But once allowed to take over the government, the Nazis put their weight behind a whole panoply of officially backed countersciences, from astrology to Himmler's "World Ice Theory," from versions of quantum mechanics that served their ideology to heinous schemes for "race purification." The readiness with which large numbers of physicians, jurists, scientists, and other academics lent themselves to the abominations committed under the last of these show that scientific literacy by itself provides no immunization; it also attests to the pliability of even so-called intellectuals when there is a cultural upheaval in which politics and parascience join. Indeed, as J. D. Bernal noted in his seminal book *The Social Function of Science,* the rise of Nazism had been prepared by irrational movements, including elements of the anti-science phenomenon in Germany at that time.[43]

In looking back on such historic cases, we can draw two important lessons. The first is that alternative sciences or parasciences by themselves may be harmless enough except as one of the opiates of the masses, but that when they are incorporated into political movements they can become a time bomb waiting to explode. We have recently been watching just such a possibility in the United States. Among the relevant documentation is an important essay by James Moore, released by the American Academy of Arts and Sciences, entitled "The Creationist Cosmos of Protestant Fundamentalism."[44] It chronicles the recent rise and political power of the anti-evolution movement in the United States. While opposition to evolutionist teachings has a long history in America, Moore notes that "today, Fundamentalists may have a fair claim that up to a quarter of the population of the US, and a rapidly increasing number of converts worldwide, live in a universe created miraculously [in six days] only a few thousand years ago, and on an earth tenanted only by those fixed organic kinds that survived a global Flood . . . The creationist cosmos of Protestant Fundamentalism has acquired an authority rivalling that of the established sciences" (p. 46).

Far from being led by old-fashioned and anti-scientist theologians of the sort familiar from the nineteenth century, the intellectual agenda of the current creationist movement has been propelled chiefly by a small but dedicated group trained in science and engineering, many with doctorates and research positions and capable of living with glaring contradictions within their total world picture. Their motivation was initially a joining of a belief in the literal truth of the Bible with a Cold War opposition to the perceived Soviet threat. They are well financed and well organized, highly productive of eloquent publications in their own journals, books, films, radio and TV programs, and educational institutes. Above all they are well connected to the most conservative political segments and church groups. Much of their activism has centered on gaining access to the minds of the young—the introduction of what they call "Scientific Creationism" into the school science curriculum through pressure exerted on local school boards—as an alternative to evolution, which they speak of as Satanically inspired and antithetical to Christianity. Beyond that, there is now indication that the movement is going to deal with Copernicus as it has with Darwin, by embarking on geocentrism.

The most noteworthy point is the joining of "Creationism" with the agenda of politically ambitious evangelists such as Jerry Falwell, Pat Robertson, Jimmy Swaggart, Jim Bakker, D. James Kennedy, and many others. "Already the shapers of opinion in church-going America . . . [despite the temporary disgrace of some] have become the most visible and influential defenders of the creationist cosmos." This movement is part of an attack on secular humanism, which they also see as part of a Satanic ideology (p. 61). As the proponents' published view shows, the stakes are much higher for them than merely displacing current biology texts. They focus on the traditional Fundamentalist task: how to prepare this world for the coming of the next.

On the way to that goal, they have encountered surprisingly little vocal opposition from the world of scholarship, science, or theology. On the contrary, they have acquired powerful allies in high places. Their sympathizers included a President of the United States in the 1980s; he is on record as holding to a world view that has open arms not only for astrology but also for UFOs, for creationism, and for a form of premillennialist Fundamentalism that concerns itself with the inevitable approach in the near future of an apocalyptic Ending. While the United States will have to live with the after-effects of many of his ideological positions, it was perhaps a lucky chance that his genial lack of deep commitment on many matters extended also to these alternative-science views and their religio-political connections; for it is sobering to think how different it might have been if he had had a driving passion for them. It may of course go the other way with some future incumbent—here or in some other country vulnerable to the same combination of forces. Moore's essay ends on the ominous note that today's *fin-de-siècle* Fundamentalism and "the reigning assumptions of liberal, evolutionary enlightenment" may yet confront each other in a *Kulturkampf*, in which they will "clash, violently perhaps, to mobilize consent and enforce political order" (p. 64).

The other lesson to draw from our historic cases is simply this. History records an important and revealing asymmetry: the original Machine Luddites of the nineteenth century were soon brutally crushed; but the Cultural Luddites have often, at least for some time, been the winners, although at great cost to their civilization. It is sobering that in every case there were intellectuals who tried to stand up to the Cultural Luddites—but they rose too late, were far too small in number, received little encouragement from their peers, and had less commitment and staying power than did their opponents.

As we have seen, history records that the serious and dedicated portion of the anti-science phenomenon, when married to political power, does signal a major cultural challenge. At its current level, this challenge may not be an irreparable threat to the modern world view as such. But it cannot be dismissed as just a distasteful annoyance either, nor only as a reminder of the failure of educators. On the contrary, the record from Ancient Greece to Fascist Germany and Stalin's U.S.S.R. to our day shows that movements to delegitimate conventional science are ever present and ready to put themselves at the service of other forces that wish to bend the course of civilization their way—for example, by the glorification of populism, folk belief, and violence, by mystification, and by an ideology that arouses rabid ethnic and nationalistic passions.

In short, it is prudent to regard the committed and politically ambitious parts of the anti-science phenomenon as a reminder of the Beast that slumbers below. When it awakens, as it has again and again over the past few centuries, and as it undoubtedly will again some day, it will make its true power known.

Notes

1. A rare example of a full-length treatment of anti-science is J. C. Burnham, *How Superstition Won and Science Lost* (New Brunswick, N.J.: Rutgers University Press, 1987). For specific facets of anti-science, see the essays by Helga Nowotny, Gernot Bohme, Otto Ullrich, and Hilary Rose in Helga Nowotny and H. Rose, *Counter-Movements in the Sciences* (Dordrecht: Reidel, 1979); and essays by Leo Marx, Lynn White, Jr., and Robert S. Morison in Gerald Holton and R. S. Morison, *Limits of Scientific Inquiry* (New York: W. W. Norton, 1979).

2. Irving Langmuir, "Pathological Science," *Physics Today,* 42 (1989): 36.

3. D. A. Bromley, "By the Year 2000: First in the World," *Report of the Federal Coordinating Council for Science, Engineering & Technology (FCCSET), Committee on Education and Human Resources* (Washington, D.C.: FCCSET, 1991).

4. J. Miller, "The Public Understanding of Science and Technology in the U.S., 1990," Draft Report to the National Science Foundation, 1 February 1991.

5. These findings are little different from those in National Commis-

sion on Excellence in Education, *A Nation at Risk: The Imperative for Educational Reform* (Washington, D.C.: National Commission on Excellence in Education, 1983). For surveys see National Science Board, *Science and Engineering Indicators* (Washington, D.C.: U.S. Government Printing Office, 1989), chap. 8; and R. G. Niemi, J. Mueller, and T. W. Smith, *Trends in Public Opinion: A Compendium of Survey Data* (New York: Greenwood Press, 1989).

6. M. Weber, "Science as a Vocation" (1918), reprinted in *Daedalus* (Winter 1958): 117 (this inaugural issue was devoted to the topic "science and the modern world view").

7. E. R. Dodds, *The Greeks and the Irrational* (Boston: Beacon Press, 1957); originally published by the University of California Press in 1951. For a more extensive discussion of it, see Gerald Holton, *The Advancement of Science, and Its Burdens* (New York: Cambridge University Press, 1986), chap. 10.

8. Niemi, Mueller, and Smith, *Trends in Public Opinion.*

9. In Richard Q. Elvee, ed., *The End of Science? Attack and Defense* (Lanham, Md.: University Press of America, 1992), p. 57.

10. S. Woolgar, ed., *Knowledge and Reflexivity: New Frontiers in the Sociology of Knowledge* (London: Sage Publications, 1988), p. 166.

11. Lionel Trilling, *Mind in the Modern World: The 1972 Jefferson Lecture in the Humanities* (New York: Viking, 1972).

12. I have treated this phenomenon in G. Holton, "Dionysians, Apollonians, and the Scientific Imagination," in *The Advancement of Science,* chap. 3. A carefully supported analysis of the fashionable attempts to link modern science and Eastern lore is S. Restivo, "Parallels and Paradoxes in Modern Physics and Eastern Mysticism," *Social Studies of Science,* 8, Part I (1978): 143–181, and 12, Part II (1982): 37–71.

13. Sandra Harding, "Why Physics Is a Bad Model for Physics," in Elvee, ed., *The End of Science?*

14. Sandra Harding, *The Science Question in Feminism* (Ithaca: Cornell University Press, 1986), p. 10; see the trenchant reviews of this issue in M. Levin, *American Scholar,* 57 (1988): 100–106, and in Clifford Geertz, "A Lab of One's Own," *New York Review of Books,* 37 (8 November 1990): 19–23.

15. Carroll W. Purcell, in M. C. La Follette and J. K. Stine, eds., *Technology and Choice* (Chicago: University of Chicago Press, 1991), p. 169.

16. Don K. Price, *America's Unwritten Constitution* (Baton Rouge: Louisiana State University Press, 1983).

17. Quoted in Clifford Geertz, *The Interpretation of Cultures* (New York: Basic Books, 1973), p. 131. Geertz elaborates the concept of world view,

as applied to a group, as follows: their "picture of the way things in sheer actuality are, their concept of nature, of self, of society . . . [including] their most comprehensive ideas of order" (p. 127); also see p. 141.

18. R. K. Merton, *Science, Technology and Society in Seventeenth-Century England* (1938; rpt., New York: Harper Torchbooks, 1970), pp. 115, 56. See Steven Shapin, "Understanding the Merton Thesis," *Isis*, 79 (1988): 594–605, for an analysis of the concept and its probable sources.

The poverty in the usage of the terms *world picture* or *world view* in English contrasts remarkably with the enthusiastic use in German, the related term *mentalité* in French, and to some degree the equivalents in Russian literature. Thus the *Oxford English Dictionary*, like most dictionaries in English, has no definition for *world picture* or *world view*, although it has an inadequate and inaccurate entry for defining *Weltbild*. For extensive definitions and discussions of the concept of *Weltbild*, see *Trübners Deutsches Wörterbuch* (1957), *Der Grosse Brockhaus*, vol. 2 (1974), and J. Grimm and W. Grimm, *Deutsches Wörterbuch* (1955), as well as K. Jaspers, *Psychologie der Weltanschauung* (Berlin: Julius Springer, 1919), chap. 2, "Weltbilder." In Russian, see entries for "Nauchnaya Kartina Mira" and "Miravazrenia," in *Filosophskii Entsiklopyedicheskii Slavar* (Moscow: Sahvyetskaya Entsiklopyedia, 1983). For a study of the *Weltbild* of an individual scientist, see Holton, *The Advancement of Science*, pp. 20–27, 57–104, and 245–248.

19. See G. Holton, "Physics Literacy," *Physics Today*, 43 (1990): 60–67. As Wittgenstein observed: "I can imagine a man who had grown up in quite special circumstances and had been taught that the earth came into being 50 years ago, and therefore believed this. We might instruct him: the earth has long . . . etc.—We should be trying to give him our picture of the world [*Weltbild*]." See L. Wittgenstein, *On Certainty* (Oxford: Basil Blackwell, 1974), p. 34e.

20. Erik H. Erikson, *Toys and Reasons: Stages in the Ritualization of Experience* (New York: W. W. Norton, 1977), pp. 147–148.

21. Peter Galison, "Aufbau / Bauhaus: Logical Positivism and Architectural Modernism," *Critical Inquiry*, 16 (1990): 709–752. For a sketch of an empiricist's personal world picture identified as "scientific humanism," see Rudolph Carnap, *Autobiography: The Philosophy of Rudolph Carnap* (La Salle, Ill.: Open Court, 1963), pp. 70–85.

22. See Holton, *Thematic Origins*.

23. M. Gerhart and A. Russell, *Metaphoric Process* (Fort Worth: Texas Christian University Press, 1984), p. 91.

24. For related studies, see the literature on the decision of the Cambridge (Massachusetts) City Council on permitting recombinant-DNA re-

search; and J. Doble and A. Richardson, "Scientific Issues and Thoughtful Public Involvement," *Technology Review,* 95 (1992): 51–54.

25. A foreign visitor might start reading on American exceptionalism by turning first to Alexis de Tocqueville's *Democracy in America* (1835–40); Robert Merton's monograph, *Science, Technology and Society in Seventeenth-Century England* (1938), and its discussion, e.g., in I. B. Cohen, *Puritanism and the Rise of Modern Science: The Merton Thesis* (New Brunswick, N.J.: Rutgers University Press, 1990); A. Hunter Dupree, *Science in the Federal Government* (Cambridge, Mass.: Harvard University Press, 1957); Daniel Bell, "The 'Hegelian Secret': Civil Society and American Exceptionalism," in *Is America Different? A New Look at American Exceptionalism,* ed. Byron Shaffer (New York: Oxford University Press, 1991); Leslie Berlowitz, D. Donoghue, and L. Menand, *America in Theory* (New York: Oxford University Press, 1988); Y. Ezrahi, *The Descent of Icarus: Science and the Transformation of Contemporary Democracy* (Cambridge, Mass.: Harvard University Press, 1990); and Gerald Holton, "The Culture of Science in the USA Today," *Methodology and Science,* 24 (1991): 55–63.

26. This danger was hinted at in Philip E. Converse's seminal article, "The Nature of Belief Systems in Mass Publics," in *Ideology and Discontent,* ed. David E. Apter (New York: The Free Press, 1964), p. 40.

27. See, for example, Isaiah Berlin, *The Crooked Timber of Humanity,* ed. Henry Hardy (London: John Murray, 1990).

28. Alex Inkeles and D. H. Smith, *Becoming Modern: Individual Change in Six Developing Countries* (Cambridge, Mass.: Harvard University Press, 1974); see also Alex Inkeles, *Exploring Individual Modernity* (New York: Columbia University Press, 1983). Other approaches stress more the economic and political definitions of modernity.

29. Of note among the other candidates is Christopher Lasch. See his new book, *The True and Only Heaven: Progress and Its Critics* (New York: W. W. Norton, 1991); see also the passionate and still timely book by C. Frankel, *The Case for Modern Man* (New York: Harper & Brothers, 1955).

30. On this last possibility, David E. Apter makes the important point: Michael Polanyi once wrote that "Men must form ideas about the material universe and must embrace definite convictions on the subject. No part of the human race has ever been known to exist without a system of such convictions, and it is clear that their absence would mean intellectual annihilation. The public must choose, therefore, either to believe in science or else in Aristotle, the Bible, Astrology or Magic. Of all such alternatives, the public of our times has, in its majority, chosen science"; see M. Polanyi, *The Logic of Liberty* (Chicago, Ill.: University of Chicago Press,

1951), pp. 57–58. But, Apter adds, this choice "embodies some very troubling and universal problems. Polanyi's statement shows a certain comfortableness about the majority choice. But what happens if, having chosen, the majority does not follow through on its choice and indeed rejects many aspects of it? What is the effect, too, upon the minority that has not made this choice—and by effect I mean particularly political consequences?" See David E. Apter, *Ideology and Discontent* (New York: The Free Press, 1964), p. 40.

31. Albert Einstein, "Autobiographical Notes," in *Albert Einstein: Philosopher-Scientist,* ed. Paul A. Schilpp (Evanston, Ill.: Library of Living Philosophers, 1949), p. 5.

32. Max Planck, *Where Is Science Going?,* trans. James Murphy (New York: W. W. Norton, 1932).

33. Kurt Vonnegut, Jr., *Wampeters, Foma & Granfaloons* (New York: Delacourt Press, 1974), pp. 163–165.

34. M. Berman, *The Reenchantment of the World* (Ithaca, N.Y.: Cornell University Press, 1981). For a brief but pungent response to the claims of what Steven Weinberg calls the "would-be sciences: astrology, precognition, 'channeling,' clairvoyance, telekinesis, creationism and their kin," see his *Dreams of a Final Theory* (New York: Pantheon Press, 1993), pp. 48–50.

35. Stephen Toulmin, *Cosmopolis* (New York: The Free Press, 1990).

36. Ironically, the rhetoric of parascience holds that on the contrary it is conventional science which is "authoritarian" and "absolute," e.g., in refusing to believe the unconfirmed and unrepeatable "first-hand reports" of people who claim to have seen, or to have been abducted by, UFOs.

37. The excerpted text was given in the *New York Times,* March 1, 1992.

38. Jan Vladislav, ed., *Václav Havel, or Living in the Truth* (London: Faber & Faber, 1987), pp. 138–139; reprinted by permission of Faber and Faber, Inc. The passage was written in 1984.

39. American Association for the Advancement of Science, *Science for All Americans* (New York: Oxford University Press, 1990).

40. See F. Gregory, *Scientific Materialism in Nineteenth-Century Germany* (Dordrecht: Reidel, 1977).

41. See, for example, M. I. Thomis, *The Luddites: Machine-Breaking in Regency England* (Hamden, Conn.: M. Archon Books, 1970); F. O. Darvall, *Popular Disturbances and Public Order in Regency England* (London: Oxford University Press, 1934); and G. Pearson, "Resistance to the Machine," in *Counter-Movements in the Sciences,* ed. Helga Nowotny and H. Rose (Dordrecht: Reidel, 1979).

42. Fritz Stern, *The Politics of Cultural Despair: A Study of the Rise of German Ideology* (Berkeley: University of California Press, 1961), p. xvii. See also the important book: Alan Beyerchen, *Scientists under Hitler: Politics in the Third Reich* (New Haven, Conn.: Yale University Press, 1977). A review of more recent scholarship is provided in Alan Beyerchen, "What We Now Know about Nazism and Science," *Social Research,* 59 (1992): 616–641.

43. J. D. Bernal, *The Social Function of Science* (London: Routledge, 1946), p. 3.

44. In M. E. Marty and R. S. Appleby, *Fundamentalisms and Society: Reclaiming the Sciences, the Family, and Education,* vol. 2 (Chicago, Ill.: University of Chicago Press, 1992).

Sources

Most chapters in this book have been substantially reworked from their original form, as given in the following publications.

Chapter 1: "Ernst Mach and the Fortunes of Positivism in America," *Isis*, 83 (1992): 27–69; © 1992 by the History of Science Society, Inc. All rights reserved. Used with permission from the University of Chicago Press.

Chapter 2: "More on Mach and Einstein," *Methodology and Science*, 22 (1989): 67–81.

Chapter 3: "Quanta, Relativity, and Rhetoric," in *Persuading Science: The Art of Scientific Rhetoric*, edited by Marcello Pera and William R. Shea (Canton, Mass.: Science History Publications, 1991), pp. 173–203. Used with permission from Watson Publishing International.

Chapter 4: "On the Jeffersonian Research Program," *Archives Internationales d'Histoire des Sciences*, 36 (December 1986): 325–336.

Chapter 5: "Spengler, Einstein and the Controversy over the End of Science," *Physis*, 27 (1991): 543–556.

Chapter 6: "How to Think about the 'Anti-Science' Phenomenon," *Public Understanding of Science*, 1 (1992): 103–128.

Index